T0213507

The Frontiers Collection

The books in this collection are devoted to challenging and open problems at the forefront of modern science and scholarship, including related philosophical debates. In contrast to typical research monographs, however, they strive to present their topics in a manner accessible also to scientifically literate non-specialists wishing to gain insight into the deeper implications and fascinating questions involved. Taken as a whole, the series reflects the need for a fundamental and interdisciplinary approach to modern science and research. Furthermore, it is intended to encourage active academics in all fields to ponder over important and perhaps controversial issues beyond their own speciality. Extending from quantum physics and relativity to entropy, consciousness, language and complex systems—the Frontiers Collection will inspire readers to push back the frontiers of their own knowledge.

More information about this series at https://link.springer.com/bookseries/5342

Masataka Watanabe

From Biological to Artificial Consciousness

Neuroscientific Insights and Progress

Translated by Tony Gonzalez

 Springer

Masataka Watanabe
School of Engineering
University of Tokyo
Tokyo, Japan

ISSN 1612-3018   ISSN 2197-6619   (electronic)
The Frontiers Collection
ISBN 978-3-030-91140-9   ISBN 978-3-030-91138-6   (eBook)
https://doi.org/10.1007/978-3-030-91138-6

Translation from the Japanese language edition: *Nou no ishiki, kikai no ishiki (Brain Consciousness, Machine Consciousness)* by Masataka Watanabe, © Chuou-kouron-shinsha 2017. Published by Chuou-kouron-shinsha. All Rights Reserved.

This Springer imprint is published by the registered company Springer Nature Switzerland AG
The registered company address is: Gewerbestrasse 11, 6330 Cham, Switzerland

Preface

Were it possible, would you upload your consciousness to a machine? What if you were soon to die otherwise? The fleeting nature of life is part of what makes it precious, but I would doubtless succumb to curiosity. If I were transplanted into a machine, what would I hear? What would I see? As time passed, would I cease to long for my previous physical form?

I believe that at some point in the future, we humans will inevitably learn how to transfer consciousness and extend our lives in some mechanical form. Today, however, realization of such a feat remains a far-off dream—we have no idea what kind of device could harbor consciousness, or indeed what consciousness even is. The science of consciousness lacks uniformity, instead remaining a confusing amalgam of varied positions. In recent years, however, a concerted effort has begun to emerge among a subset of consciousness researchers. Through their work, the vague outline of a new science of consciousness is coming into focus, revealing new paths toward a solution. The present book is an exploration of these efforts.

Simply stated, this subset of researchers—among whom I count myself—is attempting to introduce natural laws into the science of consciousness. As the term implies, "natural laws" refer to the rules by nature behaves and form the foundations of the natural world. One example is the constancy of light velocity which serves as a foundation for Einstein's theory of special relativity. Einstein's theory has revealed many strange phenomena, such as that traveling at the speed of light would stop time. It is highly probable that a phenomenon as strange as consciousness, which escapes explanation by conventional science, also arises from some set of novel universal principles.

What would such a natural law look like? To give two examples, the philosopher David Chalmers posits that all information produces consciousness, while the neuroscientist Giulio Tononi presumes that consciousness can only arise from information in a particular integrated state. I address the meaning and unique characteristics of each hypothesis in detail in the main text.

Initiatives toward introducing a natural law of consciousness aim at recasting the science of consciousness into the form that science should take. Namely, a science of consciousness should propose hypotheses based on natural laws and elucidate the nature of its subject through reproducible experiments that test those hypotheses. Doing so would finally lay bare the nature of consciousness, which for thousands of years has straddled the boundary between philosophy and science.

My own field of study is neuroscience, and I have been obsessed with consciousness for some time, just as in the words of the philosopher John Searle, "Studying the brain without studying consciousness would be like studying the stomach without studying digestion." Every day, I conduct experiments on the brains of mice, during which I constantly ponder the neural mechanisms of consciousness. And there is one thing my experience allows me to say with confidence: Throughout all of science, no subject is both so deep and so elusive as this one.

My primary goal in writing this book is to impart upon the reader the depth of this problem of consciousness. I wish to demonstrate that we need not travel to the farthest reaches of space to find problems deserving of our utmost efforts toward understanding: Such a problem resides right in our own heads. A secondary goal is to propose a method for validating the natural law of consciousness, which is indispensable to the establishment of the current approach but yet missing.

The remainder of the book is organized as follows.

Chapter 1 defines consciousness as it is addressed in this book. Specifically, I discuss "qualia," a term for subjective experiences such as seeing and hearing. I present various illusions that allow the reader to see and understand how retinal input can exist without accompanying visual qualia, and conversely how visual qualia can exist without retinal input. In many cases, however, it is difficult to comprehend that subjective experience or qualia lie at the very core of defining consciousness; hearing and seeing are all too familiar to us.

The reader need not worry, however. While qualia are the most primitive building block of consciousness, they contain the essence of all its difficulty. A thorough understanding of them lays the groundwork for understanding consciousness in general, including its most challenging aspects.

Chapters 2 and 3 relate the dramatic tale of how our predecessors established a scientific basis for consciousness. In order to understand why the natural law of consciousness is crucial, we must first understand exactly what the problem of consciousness is, and what remains unknown about it. These chapters provide that background.

We finally arrive at the true problem of consciousness in the first half of Chap. 4 and begin to see what makes it such a uniquely difficult challenge. In the words of Francis Crick—a co-discoverer of the double-helix structure of DNA who made major contributions to early understandings of consciousness—we are nothing but a pack of neurons. Contemplating just what he meant by that is essential. I expect some readers will experience quite a shock upon truly grasping what it means for consciousness to reside solely in our physical brains.

To help as many readers as possible to appreciate the true shock, I have done my best to explain, with neither deficiency nor excess, the mechanisms by which the brain operates. I describe these mechanisms at a variety of scales, from nanoscale processes at the molecular level to those involving neurons, neural networks comprising multiple neurons, regions of the brain made up of billions of neurons, and the brain as a whole. While this may feel tedious at times, my goal is not to provide a thorough understanding of these mechanisms but rather to show that there are no "black boxes" hidden within the brain. Indeed, consciousness with no black boxes is the very subject of this book.

In the second half of Chap. 4, I consider the need for a natural law of consciousness and issues related to their verification, laying out an argument for why these laws are key to taking on the challenge of consciousness from a scientific perspective.

Getting ahead of the discussion, there are limits on applying biological brains to validate such natural laws. For example, to test the above-mentioned hypothesis that information generates consciousness, we must extract only information from the brain. The biological nature of brains makes this difficult, however, because forced extraction of information kills the object of study. We may turn to analysis-by-synthesis, say, verification of natural laws through the development of machine consciousness, but it gives rise to a new problem. As of today, there is no method for testing machine consciousness. This is where the second goal of this book comes in; putting forward a test for machine consciousness.

In Chap. 5, I lay out several thought experiments regarding the proposed test for machine consciousness and examine the grounds of currently proposed natural laws of consciousness. Specifically, I question the validity of natural laws that link neural information and consciousness, and in turn suggest a natural law that involves neural algorithms.

Based on the considerations outlined above, the concluding Chap. 6 discusses future technological developments in brain-machine interface and machines that may harbor consciousness. Specifically, I put forward a new type of brain-machine interface that makes the proposed test for machine consciousness and the consequent mind uploading viable. At the start of this foreword, I mentioned that uploading human consciousness is a far-off dream, but I personally believe that day will come much sooner than many people expect.

Childhood of neuroscience is drawing to a close, and I am very excited to see what it becomes in its next stage of development and how it will change the world.

Tokyo, Japan   Masataka Watanabe

Acknowledgements

I could not have written this book without the help of many people, chief among them my mentors Shunsuke Kondo, Kazuo Furuta, Kazuyuki Aihara, Hiroshi Fujii, Minoru Tsukada, Okihide Hikosaka, Masamichi Sakagami, Ichiro Fujita, Shinsuke Shimojo, Keiji Tanaka, Cheng Kang, and Nikos Logothetis, each of whom provided invaluable resources for my research. I extend my sincerest thanks to each of them. I also thank Ryota Kanei, Naotsugu Tsuchiya, Ryusuke Hayashi, and Daw-An Wu, each of whom is younger than me, but nonetheless my seniors in terms of consciousness research. Without the thorough initiation these three provided me under the blue skies of California, I likely never would have bitten into the forbidden fruit.

I am very thankful to Tatsuya Kanbayashi, my editor of the Japanese version at the publisher Chuo Koronsha, for his never-ending patience with my always late manuscripts. It was only through his detailed advice that I was able to grasp how to write for a general audience and assemble my thoughts into a monograph. Also, I thank all the members of the Japanese reading club in Tübingen, Germany, who gave me precious feedback on the very early versions of the manuscript.

I wish to thank my high-school classmate Tomoko Yogi for the many excellent illustrations she prepared for this book. Her careful depictions helped make many otherwise confusing aspects of the brain much easier to understand.

Regarding the translation process and publishing of this English version, I greatly thank Tony Gonzalez for the original translation and countless cycles of editing, Stephen Lyle for valuable suggestions and editing after thoroughly going through the early versions, Yumiko Shibayama for also going through the early versions and assisting me with the final processes, and all the editors of Springer Nature, Angela Lahee, Emmy Lee, and Banu Dhayalan for full support.

I thank my wife Azusa from the bottom of my heart for her forbearance while I spent so much time writing this book, for her assistance with last-minute editing tasks, and for the emotional and health support she provided.

I would like to close by expressing my deepest regret at the loss of my former research partners Cheng Kang and Satohiro Tajima, both of whom passed away as I was writing this book. I am truly saddened that I never got the chance to show it to them after its completion.

Contents

Chapter 1
The Mystery of Consciousness

I Think, Therefore I Am

Presumably, you are sitting with this book in hand, and your eyes are tracing over its words. Please take a moment to look around at your surroundings. How can you be sure that the people and objects you see actually exist? Even this very book?

Even I must ask myself such questions. I see a computer monitor and keyboard in front of me, but are they really there? I hear the clicking of keys, and can feel my fingers as I type, but could these be mere illusions? Perhaps I am actually asleep in a bed somewhere, trapped in an endless dream. Or maybe my brain is connected to a computer that has me imprisoned in some virtual world. I may not even have a brain or a body; I might be nothing more than a series of electric pulses on a circuit board. You might consider this to be overly fanciful, but if it were possible to completely take control of the brain's systems for input and output, distinguishing between virtual reality and actual reality would become impossible. My situation would be much like that of the protagonist Neo in The Matrix.

Amidst all this uncertainty there is but one certain thing: my own existence. In saying so, I am not referring to my body, but rather to my consciousness, that part of me which feels as if it is viewing a monitor and typing on a keyboard. That part of me that thinks, "This is reality."

Identifying this phenomenon was a goal of René Descartes, the father of modern philosophy. Descartes began his pursuit of reality by doubting all things. Only by passing everything through the sieve of doubt, he believed, could he separate out the truth that could not be logically denied.

Visible and audible phenomena—indeed, all sensory input—were the first to be filtered out by Descartes' sieve. These could too easily be illusions or hallucinations with no basis in reality. We cannot even be sure that we are awake from one moment to the next, he argued. After all, we commonly fail to realize when we are dreaming, so believing that we are awake is no guarantee that we actually are. No matter how broad our knowledge, if we cannot trust our own senses, then reality stands on a shaky foundation.

© Springer Nature Switzerland AG 2022
M. Watanabe, *From Biological to Artificial Consciousness*, The Frontiers Collection,
https://doi.org/10.1007/978-3-030-91138-6_1

In this way, Descartes removed all doubtful things one by one, until in the end he was left with a single phenomenon that he could not exclude: his own mind, actively attempting to exclude all other things. In other words, his consciousness. From this, he derived one of the most famous lines in philosophy: "I think, therefore I am."

This Cartesian "I" is the focus of this book. It is our starting point for inquiring into consciousness. It is the one undeniable aspect of existence. Even if we are merely brains in jars, even if we are just programs running on some advanced computer, when we reflect on our own existence, there is no doubt that we exist.

However, this "I" alone does not provide a sufficient basis for scientific inquiry into its nature. If we doubt everything beyond our own existence, then even our experimental equipment resides in a fog of uncertainty. Indeed, we know of no technique external to our own brains for establishing the "self." Hence, if we hope to subject consciousness to scientific scrutiny, we must maintain "self" as our primary object of study, yet capture it in such a way as to allow verification through observation and experimentation.

Subjective Experience: The Ultimate Reduction of Consciousness

To unravel the mechanisms of consciousness, we must first reduce it to its most basic elements. Most people would agree that today's computers do not exhibit consciousness. But what is the essential difference between them and us?

Of course, computers have achieved some remarkable feats. In 1997, the IBM supercomputer Deep Blue defeated world chess champion Garry Kasparov. In 2017, the AlphaGo program, developed by a Google subsidiary, bested Ke Jie, the highest-ranked go player at the time. The problem-solving abilities of computers have surpassed those of humans in many other fields with well-defined sets of rules as well. Through "deep learning," the latest development in machine learning, software programs are even surpassing humans in areas previously considered beyond their reach, such as image recognition.

You have probably seen images like that in Fig. 1.1, a string of CAPTCHA text distorted to purposely impair readability. These are used to prevent automated creation of accounts for online services, on the basis that humans, but not computers,

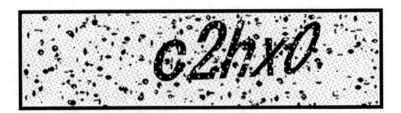

Fig. 1.1 CAPTCHA text

can read them. Don't expect these to be around much longer, though; computers are learning to read them even better than humans can.

Despite such remarkable advancements, however, there is one thing that computers don't have, and that some scientists and philosophers think they never will: subjective experience of vision, audition, tactile sensation, emotion, decision making, etc. In other words, qualia.

If you have heard of qualia before, seeing this word may set you on edge. Definitions found online are typically vague, or even borderline incomprehensible. In truth, however, the definition of qualia is not all that complex. In terms of vision, it simply means "the experience of seeing." You can see a face in front of you, but you cannot see one behind you. There are visual qualia in what you see, and none in what you cannot. That's all.

The hard part is the "qualia problem." Namely, why do qualia, or subjective experiences, occur in beings with brains, and only in beings with brains? Modern digital cameras can capture a scene, search for faces within that scene, and adjust their focus accordingly. But the camera does not "see" the scene or the faces. Digital cameras do not experience visual qualia.

Getting a feel for this fact is the first step toward understanding the qualia problem. We are so used to seeing the world that we easily fall into the trap of assuming that cameras see it too. The important thing here is the fundamental difference between processing and recording an image and seeing the world. Because this is the first potential stumbling point, let's consider it more carefully through some examples.

Allow me to repeat, however, that this is only the first step toward understanding the qualia problem, the heart of which is its mere existence in our minds. To fully understand that, we must learn about brain mechanisms, along with the progress of consciousness research up until now. Doing so will lead us to the surprising fact that our brains are, in essence, nothing more than electronic circuits, not critically different from a digital camera. If so, given that cameras cannot "see," why can we? After some preparation, we will revisit the full complexity of the qualia problem in Chap. 4.

The Fiction of the Visual World

Qualia are such matter-of-fact phenomena to us that it is difficult to comprehend that they only exist in conscious beings. Such comprehension requires a certain transformation of concepts, starting with the fact that the world we see is completely unlike the world as it actually is. We do not observe the world itself—what we see instead is our brain's interpretation of the world, a creation based on visual information from the eye.

For example, every day, we experience a visual world that we consider being full of natural color, but color is not an aspect of the real world—it is a fabrication of the brain. The actual world is a dreary place filled not with color, but with electromagnetic waves. These waves—which allow you to listen to the radio, watch television, heat food in your microwave, and see—are all in essence the same thing. The only difference is how long their wavelengths are. The light we can see has wavelengths between approximately 4/10,000 and 8/10,000 of a millimeter (400–800 nm). Shorter and longer wavelengths are invisible to us.

Interestingly, we perceive red and purple as similar colors, but as evidenced by the words "infrared" (wavelengths too long to be seen) and "ultraviolet" (those too short to be seen), these colors lie on opposite ends of the visual spectrum. So while the physical properties of these colors are very different, we perceive them as being similar (see Box 1 for its possible reasons). Hence, qualia experienced when viewing colors must be a creation of the brain.

It may seem as though your eyes scan the three-dimensional world like searchlights in the dark, directly accessing and revealing the world as it is, but that is not the case. Rather, your brain combines two sets of imperfect visual information (3D to 2D compression, drastic loss of resolution in peripheral vision, etc.) and compiles the results for "you." Your brain so cunningly creates this virtual world that it seems real and authentic.

Box 1: Retinal Photoreceptors and Color Perception

Much of our ability to sense light as color comes from special cells (cone cells) embedded in our retinas. There are three types of cone cells (Fig. 1.2, top), which react to the wavelengths of the three primary colors that we perceive: red, green, and blue (Fig. 1.2, bottom). Each type of cone cell reacts more strongly to its preferred wavelength than the other two types. For example, the peak reaction in a "red" cone cell occurs at around 630 nanometers (one nanometer is one-billionth of a meter), but "green" cone cells react only weakly to those wavelengths.

The reds and purples that mark the lower and upper limits of visible light may seem similar to us because, at these extremes, none of the three types of cone cells have a strong reaction. So while the physical properties of these wavelengths are very different, our brains perceive them in similar ways.

One other note about the three primary colors: We perceive a combination of red and green light as yellow because cone cell activation at those wavelengths (680 and 550 nm) is similar to that resulting from activation due to a single intermediate wavelength (620 nm; Fig. 1.2, bottom). Given that the responses are analogous at the retinal level, our brains situated downstream cannot distinguish between the two stimulus configurations (680 + 550 nm vs. 620 nm alone). In this manner, the three primary colors of light and light synthesis rules such as "red + green = yellow" or "green + blue = cyan" are totally a creation of our brain, following the property of our retinas.

Note that many mammals have only two types of cone cells, and therefore experience two, not three, primary colors. They are able to distinguish between far fewer types of light than humans, and do not perceive light synthesis rules

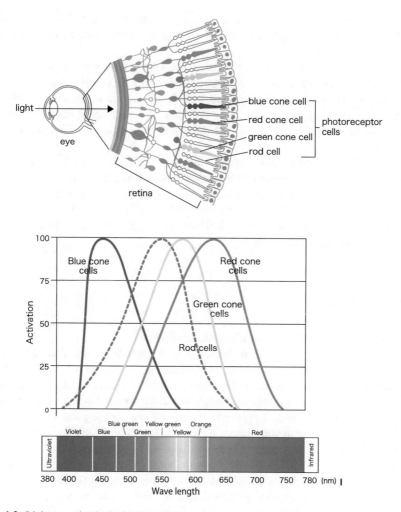

Fig. 1.2 Light reception in the human retina

like we do. The world of color that they live in must therefore be very different from ours.

Imagine Being a Mole

To better understand how our brains compile a virtual visual world that does not necessarily match reality, imagine you have turned into a mole. (We will assume for now that you-as-mole is a conscious entity.) Considering that the structure of a mole's eyes is less developed than that of a human's, we can assume that your visual world is very fuzzy.

Say you are crawling in your dim tunnel, toward a faint brightness ahead, and a dark shadow crawls across your field of vision. It's an earthworm, one of your favorite snacks. You grasp it with your well-developed claws, through which you confirm other details, such as its slimy skin and long, segmented body. As a mole, you might think, "This is the real world, quite unlike the nebulous blur that I see with my eyes."

What is important here is the truth of the qualia behind all this. The fuzzy scene your mole eyes see is nothing like the actual world; it is a pure fabrication of your tiny mole brain. Even so, that blurry visual world, no matter how unlike reality it may be, is a quale ("quale" is the singular form of "qualia"). You might not realize the difference between quale and reality if you were a mole, but it is worth considering while you imagine you-as-mole.

Seeing Things

Let's consider the matter from a different perspective.

First, look at Fig. 1.3, which demonstrates a phenomenon called neon color spreading. You probably see a faint, semitransparent square in the middle, but no square actually exists—the image shows only concentric circles with quarter sections of different brightness. But even knowing that, the square persists as an incontrovertible creation of your brain. To prove it to yourself, try covering the two upper sets of concentric circles with your hands. As soon as you do, the central top edge of the square that was so clearly present before disappears into the page.

Your brain shows you this phantom square because it would be "natural" for it to be there. This raises the question of what "natural" means in this context. The answer to that is closely connected to the "unnaturalness" inherent in the figure.

The actual stimulus is an arrangement of twelve circles arranged into four sets of three concentric circles each. In each of these sets of grayscale circles, a 90° segment has a different luminance. Take a look at just one of the sets of concentric circles. As you can see, the boundaries marking the change in the circles' luminance form perfectly straight lines. The lines are too straight, in fact—were you looking at, say, a piece of rusty metal, you would never expect such a perfect linear alignment of pattern to occur by mere chance. Even more unnatural is the perfect alignment of the luminance boundaries between the four sets of concentric circles.

In what type of situation would you find such perfect alignment? The simplest case would be that a single, semitransparent square is placed on top of the concentric circles. If we interpret things in this way, then of course we expect a perfect alignment of luminance boundaries. Thus, our brains present such an image to us.

Your brain strives to provide you with the most faithful visual perception possible, but at the same time, it does its best to show you a natural interpretation of the world.

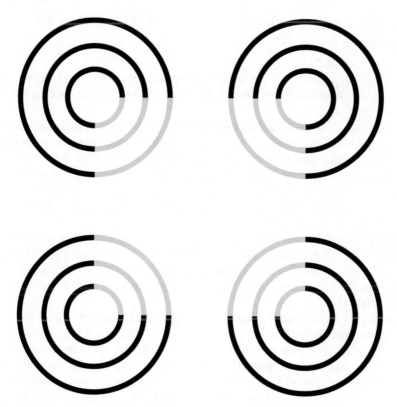

Fig. 1.3 Neon color spreading

In this tug-of-war between faithfulness and naturalness, strange things can happen; it fabricates things that do not actually exist. Furthermore, the brain's calculations of naturalness are not simple; they involve many factors, such as comparisons of distant visual objects. So while the visual world is a creation of our brains, that creation is the result of highly complex visual processing.

Moreover, we experience perfectly fine qualia even when dreaming. In fact, some experimental studies suggest that if we consider only the quality of sensory experiences, dream qualia can be equal to or superior to those experienced while awake. This is true despite the fact that the sleeping brain is largely isolated from the outside world and the body. The dream world—a world wholly segregated from the reality of lying in bed—is, without a doubt, a fabrication of the brain. If that is the case, then surely it is not difficult for our brains to create subjective experience with the aid of our eyes and ears when we are awake.

In contrast, today's computers do not dream, and do not possess subjective experience. To answer the question, Philip K. Dick posed in his 1968 novel that androids do not (yet) dream of electric sheep.

Visual Processing in the Human Brain Without Subjective Experience

Allow me to introduce a counterexample: patient DB, who, following surgery, developed the curious ability to carry out visual processing unaccompanied by subjective experience.

When DB was 26 years old, as a treatment for a brain tumor, he underwent surgery to remove a brain region called the primary visual cortex, the primary entry point for visual information to the cerebral cortex. Losing it left DB subjectively blind. The interesting thing, however, is that when asked to describe the location or movements of nearby objects, he was able to do so with surprising accuracy. Even DB himself was unable to explain this ability. Somehow DB's brain was performing visual processing without subjective experience.

This phenomenon was later named "blindsight." While the term may seem self-contradictory, it describes blindness in a first-person sense, accompanied by sight in a third-person sense. It effectively demonstrates that subjective experience exists only when the brain successfully creates them, not as an automatic byproduct of visual processing.

Next, let's experience for ourselves retinal input that does not accompany subjective experience.

In everyday life, the view from each of your eyes differs due to the space between them, but each nevertheless provides nearly the same image. The two similar images are fused in our brains to create a 3D percept. Meanwhile, when each eye is presented with a completely different image, they compete for dominance of your subjective visual experience, a phenomenon called binocular rivalry. This occurs, for example, when the horizontal pattern in Fig. 1.4 is presented to one eye, and the vertical pattern is presented to the other.

To experience this yourself, position the figure so that each of the two patterns are viewed by a different eye. This will occur when the images are placed at just the right distance from your eyes, which will depend on which viewing method (see caption) you use. At this point, they would overlap in your field of vision, and something interesting occurs: you should see vertical lines, then horizontal lines, then back to vertical in a sequence that switches every few seconds. Except for the very short time during which the switching occurs, there is very little confusion between the two percepts.

The critical point here is that when you are seeing the vertical lines, you don't see the horizontal lines at all, despite the fact that both images are surely entering your brain. This invisible stimulus during binocular rivalry is a prime example of visual input that does not result in subjective experience. We will elaborate on it in the next chapter.

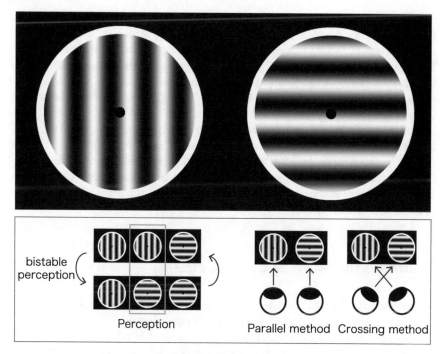

Fig. 1.4 Binocular rivalry. In the parallel method, the viewer looks far away without focusing, while in the crossed method, the viewer focuses on a finger held up between the page and the eyes. Adjust the distance to the page until the image appearing in the middle cleanly overlaps (lower left image)

Subjective Experience Is a Privilege of the Conscious

I hope the preceding discussion has made clear that subjective experience is not some kind of extra bonus occurring as a side-effect of visual processing. Subjective experience may arise without physical support (e.g., neon-color-spreading), whereas physical existence does not necessarily lead to subjective experience (e.g., binocular rivalry). Subjective experience is a privilege of the conscious, and indeed is the essence of consciousness.

Of course, I'm sure that some readers will remain unconvinced. The experience of seeing and hearing is all too common, you might think. Surely the essence of consciousness is something more advanced, more complex than that? As a conscious being, it is no surprise that you might think that. However, nearly all researchers in the field agree that the aspect of consciousness that makes it so difficult to address is all contained in the problem of subjective experience. I discuss the true nature of the problem in Chap. 4, by which point I believe you will be convinced.

By the way, subjective experience is not limited to the five senses. Other types include ones resulting from thoughts and memories. I mentioned the amazing chess abilities of computers above, but computers surely do not experience the same sensations that humans do when pondering their next move, nor the sensation associated with the flash of insight that reveals the perfect response to an opponent's move. When performing image recognition tasks, computers do not experience the sensation that humans do when finally recalling a forgotten name as they look at an old photograph. In each case, these sensations are subjective experiences—the primary focus of consciousness science, and of this book.

Thus our definition of consciousness is simple; "what we experience all the time," in other words, subjective experience. The true difficulty lies in associating these subjective experiences with the brain. Specifically, how and where do they arise? Grasping the perplexity of that question requires some knowledge of the brain, starting with how neurons work. By what mechanism could the neurons in your brain possibly make up "you"?

Neurons as the Source of Consciousness

As mentioned in the introduction, early consciousness researcher and co-discoverer of DNA's double-helix structure, Francis Crick once claimed, "You are nothing but a pack of neurons." Neurons are cells with nuclei, just like all other cells that make up your body, that comprise the primary anatomical unit in the brain (Fig. 1.5).

Modern neuroscience research has left little doubt that "you" are a product of neural activity. But what did Crick mean by "nothing but"? To understand, we must

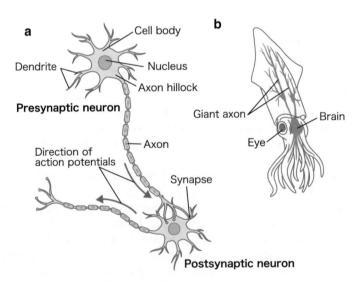

Fig. 1.5 A typical neuron and the giant axon in a squid

peel back the mystic veil of brain processes, starting off from the inner workings of neurons.

The brain remains the source of many mysteries, but the behavior of its individual neurons is not one of them. Pioneering research by Alan Hodgkin and Andrew Huxley, recipients of the Nobel Prize in Physiology or Medicine for their eponymous equations, provided a clear picture of the critical mechanisms by which neurons operate from the molecular to the cellular level. I therefore begin by retracing Hodgkin and Huxley's steps in determining how neurons work.

The following description is somewhat detailed, but it is worth being exposed to these processes to fully comprehend that neurons have no tricks hidden up their sleeves. If my explanation becomes too complicated, feel free to skim. Only the gist of what follows will be needed later, and I will provide summaries as required.

Research by Hodgkin and Huxley

Andrew Huxley graduated from the University of Cambridge in 1938, while Alan Hodgkin was an up-and-coming neuroscientist at the Marine Biological Association Laboratory in Plymouth. The two began working together on the inner mechanisms of neurons in that seaside town in southern England as Hodgkin invited Huxley to join his group. Neurons are distinguished from other mechanisms in a living body by their unmatched ability to rapidly transmit information, and the pair hoped to isolate the unique structures that made this possible.

Neurons primarily comprise three components: dendrites, which accept input, a cell body (or "soma") that compiles that input, and an axon, which performs output (Fig. 1.5a). At the time, it was known that some kind of electric process was occurring within neurons, but the details remained a mystery because the technology of the time did not allow for direct observations. Huxley and Hodgkin, for the first time in scientific history, managed to insert fine electrodes into neuronal axons, which allowed them to directly measure the electric potential within them. (We will see later why they chose axons, not soma or dendrites, for their measurements.)

Here, an electric potential is something like a difference in altitude in a world of electricity. For example, the positive terminal in a 1.5V battery has a 1.5V potential with respect to the negative terminal. Just as water flows from higher to lower altitudes, electric current flows from positions of higher to lower potentials.

Taking advantage of their seaside laboratory, Hodgkin and Huxley chose the giant axon in squid (Fig. 1.5b) for their studies. Squid swim by filling their body with water and rapidly expelling it behind them, and a neural fiber called the giant axon controls the structure by which they do so. While axons in a human brain have diameters of only a few tens of microns, the giant axon in a squid neuron measures up to one millimeter. But, a millimeter is still just a millimeter, so inserting an electrode was no easy task. The two went on to develop methods for observation of electrode insertion angles, and finally succeeded in capturing electric potential changes within axons (Fig. 1.6a).

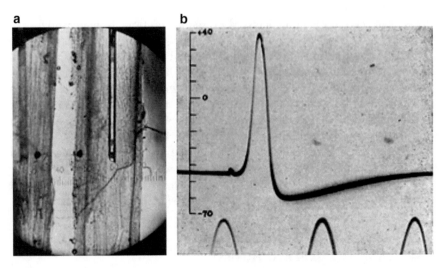

Fig. 1.6 a An electrode inserted into the giant axon of a squid, and **b** an action potential measured by the electrode (Reprinted from Hodgkin and Huxley 1939)

Here is what they found. Normally, axons retain a lower potential in their interior than in their exterior. In the diagram in Fig. 1.6b, this is shown by the negative values at the left. Measuring changes in potential showed a sudden increase followed by an equally rapid descent and return to the normal state. This rise and fall lasted for only 1/1,000 of a second. What Hodgkin and Huxley had captured were the "action potentials" that travel throughout the brain to pass on information.

Several weeks after their successful measurements, Germany invaded Poland, triggering the Second World War and interrupting their research. The two hurriedly compiled what they had discovered and published a paper titled "Action potentials recorded from inside a nerve fibre" in the 21 October 1939 issue of Nature. Allow me to admire the simple, direct title of the paper, something we don't often see in neuroscience today.

After this, the two dedicated themselves to wartime research. Hodgkin worked on developing an oxygen mask for pilots, and later became involved in radar research. Huxley worked on applying control theory to improve the accuracy of machine guns. As it turned out, these wartime experiences greatly contributed to their later studies.

Hodgkin and Huxley teamed up again six years later after the war was over. Armed with two new tools, they took on the challenge of determining the biological mechanisms that produce the action potentials they had discovered.

Their first tool was a device that forced axon interiors to instantaneously switch to a given potential. This device allowed them to measure electric currents flowing in and out of axons when the potential was switched to a new value. Their goal was to divide the mechanisms by which action potentials appeared (rising potential, falling potential, etc.) and clarify the mechanisms operating at each stage. As an analogy, consider foreign exchange markets. Currencies rise and fall all the time, but the

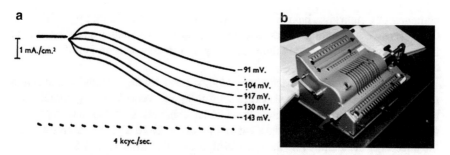

Fig. 1.7 **a** Current flowing in and out of an axon when its potential is changed from one value to another, and **b** Huxley's manual calculator (**a** Reprinted from Hodgkin and Huxley 1952, **b** Reprinted from Schwiening 2012)

factors behind devaluation of the yen when the exchange rate is 80 yen per dollar are very different from those when the rate is 150 yen per dollar. In the former, case the Japanese government may have implemented an economic stimulus package, while in the latter, the economy may have taken a downturn. While impossible in practice, repeatedly performing social experiments with the dollar–yen exchange rate would allow us to isolate factors behind currency fluctuations.

This is very similar to what Hodgkin and Huxley were after with the mechanisms of action potential generation. Figure 1.7a shows changes in current when the internal potential of an axon was switched from one value to another. These measurements, combined with the pair's second tool, led to a historic discovery.

Cranking Out Hypotheses

The pair's second tool was a mathematical method that Huxley had acquired during the war.

First, they listed hypotheses for the mechanism behind action potential generation, then derived mathematical formulas representing each one. Comparing the results predicted by those formulas with actual experimental results allowed them to identify the correct formula—that is, the correct mechanism.

Hodgkin and Huxley hypothesized that axonal surfaces are covered with countless structures called ion channels, and that these are the biological mechanism by which action potentials occur. An ion is a particle with an electric charge, and an ion channel acts as a pathway that allows ions to pass between the interior and exterior of an axon.

We know today that there are many types of ion channels, but Hodgkin and Huxley were particularly interested in the type that opens according to the electric potential within the axon. They conjectured various characteristics regarding these ion channels and represented each as a mathematical formula.

These formulae were simultaneous differential equations, and thus, impractical to solve using pencil and paper alone. Their original plan was to use the EDSAC I computer at Cambridge, but unfortunately, the computer was being upgraded at the time. They were thus forced to use a manual calculator that Huxley had lying around in his office (Fig. 1.7b), which reportedly required three weeks of literally cranking out calculations.

As a result of this laborious analysis, Hodgkin and Huxley concluded that action potentials occur through a well-coordinated relay of events involving mainly two types of channels: sodium ion and potassium ion channels. Both ions have a positive charge (which raises the potential), where the former has a higher concentration on the outside of the axon, while the latter is higher on the inside (Fig. 1.8).

When a certain trigger (more on this below) slightly increases the potential within an axon, many of its sodium ion channels start to open, allowing sodium ions to flood into the axon, immediately raising its potential. These sodium ion channels then start to close, and at nearly the same time, the axon's potassium ion channels start to open. This stops the inflow of sodium ions and creates an outflow of potassium ions, causing the elevated potential within the axon to immediately fall. This delicately timed opening and closing of two types of ion channels are what creates action potentials (see Box 2 for a more detailed description).

One thing to keep in mind is that these mechanisms also apply to the initial generation of an action potential within a neuron. Please take one more look at Fig. 1.5a. We know today that action potential generation begins at a site called the "axon initial segment," situated at the connection between the soma and the axon. Countless voltage-gated sodium and potassium ion channels are located at this connection point, awaiting the trigger (a slight increase in potential; details described later) for an action potential.

Box 2: Action Potential Generation in Axons

Let's take a closer look at the mechanisms that Hodgkin and Huxley's equations revealed. The equation involves three types of ions, sodium, potassium and chloride, but for simplicity, we focus on the first two.

Channels that allow the passage of sodium ions consist of four-ply "gates" (Fig. 1.9). Three of these gates ("m" in the figure) open wider as the potential within the axon increases, while the fourth ("h" in the figure) closes as the potential increases.

Potassium ion channels similarly have four gates ("n" in the figure), and all act in the same way, opening wider as the potential within the axon increases. Another critical difference is that these gates open more slowly than those in the sodium ion channels. In both cases, ions cannot pass until all four gates are open.

Action potentials are generated through the coordination of both types of ion channels. In the initial state, a negative potential is retained within the axon. When a certain trigger (described later) slightly increases the potential, three of the closed m-gates in some sodium ion channels start to open, whereas

1) Three "doors" of sodium channels open up and sodium ions start
flowing in = <u>Axonal potential increases</u>

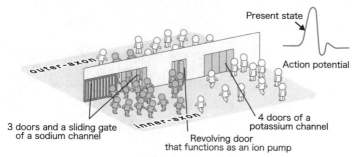

2) Increase of membrane potential leads to closing of the sodium channel
sliding gate = <u>Axonal potential saturates</u>

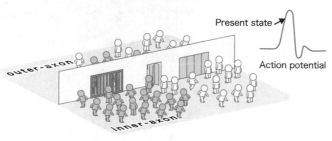

3) All four doors of potassium channels open up and potassium ions
start flowing out = <u>Axonal potential decreases</u>

Fig. 1.8 Voltage dependence: the mechanism by which ion channels produce action potentials

the remaining h-gate is already open. So some sodium ion channels will then
have all four gates open, allowing sodium ions to flow in (Fig. 1.8, top). This
raises the potential within the axon, causing the three gates in other sodium ion

Fig. 1.9 Voltage-gated ion
channels

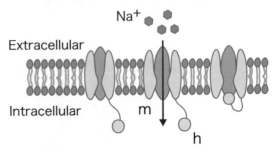

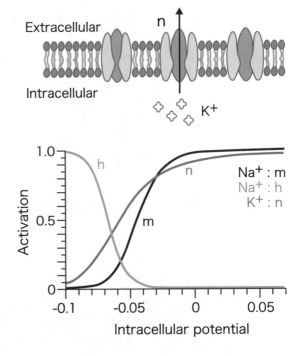

channels to open as well, in turn causing an even more rapid inflow of sodium
ions. This sharply raises the potential in an avalanche-like manner.

As the potential increases, the h-gate in the sodium ion channel begins to
hit the brakes. This gate closes as the potential within the axon increases, so at
a certain point, sodium ions can no longer enter (Fig. 1.8, middle). Recall that
ions cannot pass if even one gate in the ion channel is closed.

Fig. 1.10 Actual structure of a voltage-gated ion channel (Reprinted from Jiang et al. 2003)

This stops the increase in potential, but how does it drop? This is where the potassium ion channels come into play. As mentioned above, all four of the n-gates in potassium ion channels open wider as the potential increases, similar to the m-gates in potassium ion channels. The difference is that they do so more slowly. So around the time the potential peaks out, these gates open, allowing potassium ions to flow outside the axon (Fig. 1.8, bottom). This lowers the potential within the axon, completing the generation of an action potential. This time-lagged operation of sodium and potassium ion channels allows for sharp potential increases and decreases that last only 1/1,000 of a second.

When Hodgkin and Huxley took on the challenge of examining the mechanisms behind action potentials, not only that the structures of ion channels were unknown, but also, their existence was totally hypothetical. The details of the potential mechanism existed only within their heads, and the only hint that they reflected the actual biological mechanisms lay in the agreement between the results of their differential equations and their experimental observations.

Fifty years after their publication, experimental proof was provided that their proposed mechanism was correct: in the 1990s it finally became possible to directly observe ion channel structures, revealing the 3 m-gates and a single h-gate of sodium channels and the 4 n-gates in potassium channels, each constructed through exquisite combinations of proteins (Fig. 1.10). Furthermore, cutting-edge observational methods have demonstrated that the various ion channels operate exactly as Hodgkin and Huxley predicted.

It is remarkable that just one set of equations describes the fundamental nature of information propagation that makes a brain a brain, confirmation of which required decades of research. Hodgkin and Huxley produced truly monumental research that demonstrates the power of applying mathematical methods to biology.

Are Neurons Connected? The Debate Between Golgi and Cajal

The above discussion has provided an overview of the mechanisms behind action potential production in neurons: action potentials are produced in the axon initial segments of neurons and propagate along axons to the next neuron. Here the propagation of action potentials along axons occurs by way of a chain reaction like a domino falling, where a rise of potential caused by a neighboring action potential triggers the next action potential, all the way up to the next neuron. This raises the question of how that next neuron is connected. Are they in direct contact?

The Italian scientist Camillo Golgi was the first to hypothesize an answer to this question through observation. To do so, he sliced brains very thinly and attempted to examine them under a microscope. Unfortunately, such simple observations were insufficient to allow neurons to be seen. Even thin brain slices contain many layers of neurons, which are furthermore semitransparent, making individual neurons impossible to observe.

To resolve this problem, Golgi developed a method for staining neurons using silver nitrate and potassium dichromate. This form of staining later came to be known as "Golgi's method." One interesting feature of the Golgi method is that it does not stain most neurons, but those that it does are stained in their entirety. Why this happens remains unknown, but what is certain is that it brings out just the right number of neurons under the microscope. Using this method, Golgi carefully searched for attached neurons that happened to be revealed in the staining process. Through these observations he determined that neurons are directly connected, supporting the "reticular theory" originally postulated by the German anatomist Joseph von Gerlach in 1871.

The prominent Spanish anatomist Santiago Ramón y Cajal disagreed with Golgi's conclusions, proposing instead that minuscule gaps exist between neurons, the so-called "neuron doctrine." Ironically, Cajal was an ardent proponent of Golgi's method. He prepared brain samples using Golgi staining and recorded them in meticulous sketches. At the time, some methods had been developed for photographing images from microscopes, but they could not capture fine details. Both Golgi and Cajal thus had to rely on their own eyes and report their observations using hand-drawn sketches. Comparing their sketches, we can see that those created by Golgi (Fig. 1.11, upper left) do indeed show directly connected neurons, while Cajal's show a small inter-neuron gap (Fig. 1.11, upper right). So who was right?

In fact, normal optical (light-based) microscopes have physical limitations in spatial resolution, and even the most advanced optical microscopes of today lack the performance to clearly answer this question. The lower left image in Fig. 1.11 shows a modern optical micrograph of a brain sample stained using the Golgi method. Its lack of clarity clearly demonstrates that Golgi and Cajal had to rely on their imagination when depicting connections between neurons.

In 1906, both received the Nobel Prize in Physiology or Medicine in recognition of their work on the structure of the nervous system. At that point, however, their debate remained unsettled. According to an anecdote, they refused to even look each other in the eye during the award ceremony.

What finally settled the matter was the invention of the electron microscope. As our current use of the word "neuron" suggests, Cajal's neuron doctrine won out. Close examination of the connections in the electron micrograph in Fig. 1.11 (lower right) shows separations, albeit tiny ones: the gaps are only around 2/100,000 mm (20 nm) wide. Standard optical microscopes have resolutions of only around 2/10,000 mm, a tenth of the resolution that would be necessary to reveal the gaps, so it is no wonder that Golgi and Cajal came to opposing conclusions.

For the following reference, the connection between neurons is called a "synapse," and the gap between them is called the "synaptic cleft."

Information Transmission Between Neurons

While Golgi and Cajal continued to debate, another argument was brewing over the connections between neurons: the so-called "soup versus spark" debate regarding how signals propagate between neurons. Specifically, it was unclear whether signals were transmitted as electricity ("sparks") or as chemical reactions in a medium ("soup," namely cerebrospinal fluid).

It was the German pharmacologist Otto Loewi who settled this debate through an ingenious experiment using two frog hearts.

Two nervous systems control the beating of hearts: "sympathetic nerves" make the pulse beat more rapidly, while "vagus (parasympathetic) nerves" make it beat more slowly. Loewi first applied electric currents to activate vagus nerves leading to the first heart (Fig. 1.12). Due to the direct effect of nerves, this lowered the heart's

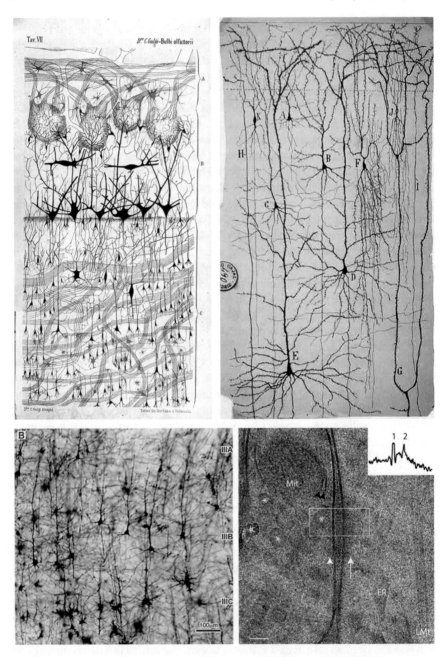

Fig. 1.11 (Upper left) Sketch of an olfactory bulb by Golgi (Reprinted from Golgi 1875). (Upper right) Sketch of the cerebral cortex by Cajal (Reprinted from Cajal 1904). (Lower left) Optical micrograph of the cerebral cortex (Reprinted from Džaja et al. 2014). (Lower right) Electron micrograph of the synaptic cleft. Scale bar at left bottom spans 50 nm (Reprinted from Zuber et al. 2005, Copyright (2005) National Academy of Sciences, U.S.A.)

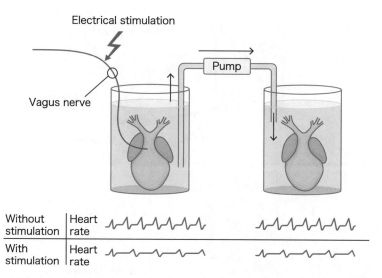

Fig. 1.12 The experiment by Otto Loewi

pulse. He next circulated the fluid that the first heart was submerged into a vessel containing a second heart. Upon doing so, the second heart also started beating more slowly.

The two hearts were electrically shielded, so the only thing they shared was the fluid they were bathed in. The sole explanation for the slowdown, therefore, was that electric stimulation of the vagus nerves in the first heart produced some chemical, which was passed on to the second heart. Loewi therefore concluded that the "soup" carries signals over the final gap.

Loewi named this unknown chemical *Vagusstoff*, which is German for "the substance of the vagus." The actual chemical was later identified and named acetylcholine.

Neuronal Interaction Through Neurotransmitters

Chemicals that bridge the gap between neurons are called neurotransmitters. Today, we know that the acetylcholine discovered by Loewi is but one of over fifty known neurotransmitters. Another is glutamic acid, a component of the popular seasoning MSG.

Information travels between neurons via neurotransmitters as follows. First, suppose that an action potential is traveling along an axon in neuron A, heading for neuron B (Fig. 1.13, top). When the action potential reaches a synapse, neuro-

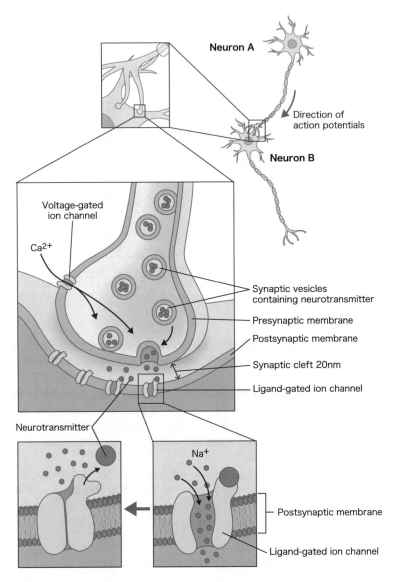

Fig. 1.13 Neurotransmitters and neurotransmitter-gated ion channels transmit signals across a synaptic cleft

transmitters are released into the synaptic cleft. The neurotransmitters spread into the cleft, and some reach neuron B (Fig. 1.13, middle).

Something interesting happens next: the neurotransmitter reaching neuron B acts as a key that opens a special kind of ion channel (Fig. 1.13, bottom). These channels are located on the surface of neuron B's dendrites and are different from that discovered by Hodgkin and Huxley in that their opening and closing do not rely on electric potential; the ion channel simply opens when the correct neurotransmitter key arrives, and it remains closed otherwise (Fig. 1.13, bottom). An open ion channel allows specific ions to flow in or out, with the direction determined by the balance between external and internal ion concentrations. Eventually, the ion flow results in a potential change of neuron B (called "postsynaptic potential"), which can be either positive or negative, depending on the charge and flow direction of ions.

Neuron Thresholds

We have come to the point where an action potential occurring in neuron A has triggered a change in potential in neuron B. What's left to see is how this change in potential produces an action potential in neuron B.

Synapses are concentrated in a part of the neuron called dendrites (Fig. 1.14). A potential change occurring in one synapse spreads through these dendrites and eventually reaches the neuron's soma. The soma takes in these potential changes from the neuron's many dendritic branches, and subsequently, these values are aggregated as a single sum potential. (lower graph in Fig. 1.14).

This, finally, is the trigger that the axon initial segment has been waiting for. The axon initial segment is tightly connected to the soma, so both have equivalent values of electric potential. When the above sum potential reaches a certain value

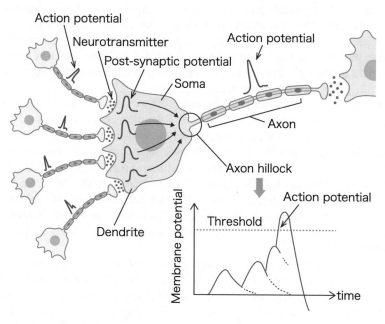

Fig. 1.14 Neuron thresholds

(the threshold in the graph), voltage-dependent ion channels discovered by Hodgkin and Huxley kick into action, and an action potential is triggered at the axon initial segment. To sum up, individual neurons function as a thresholding unit, where an action potential is emitted when the total input from other neurons exceeds a certain value.

Let's next turn to the actual operating point of a neuron in the brain. An average neuron accepts synaptic input from thousands of other neurons. Since a single neuron can produce anywhere from several to even hundreds of action potentials per second, the total input count sums up to tens of thousands per second. Each arriving action potential traverses a synapse as neurotransmitters and is converted into either a positive or negative change in potential. We can think of this as similar to stepping on the gas pedal and the brakes at the same time while driving. If the gas pedal is pressed down even a little bit more than the brake, then the neuron emits an action potential (called "neuronal firing").

The Brain Learns by Fiddling with Synapses

To accomplish signal transmission between neurons, the brain goes from action potential to neurotransmitter, then neurotransmitter to change in electric potential. But why would evolution lead to such a roundabout method? Wouldn't it be simpler to directly transmit electric signals? Simpler, perhaps, but this would place harsh constraints on how the brain processes information. Specifically, the only information that can be transmitted via an action potential is whether the signal is on or off. If our brains were to use action potentials to directly transmit information between neurons, then that information would be limited to the dull binary of ones and zeros. This seems far too rigid of a functional architecture for our brains, and would clearly limit its capacity for information processing.

Synapses solve this problem by converting binary ones and zeros to continuous negative or positive values. Furthermore, as I describe below, neurons can adjust these values according to their destination to accomplish the desired processing.

An interesting feature of our brain is that the number of neurons has not changed much from the time we were born, yet we were able to learn plenty since. By "learning" I do not mean just book learning, but learning in a broader sense: new faces, new streets, even how to ride a unicycle. The fact that the number of neurons basically stays constant suggests that acquiring new knowledge, new memories, and new motor skills do not require new neurons. Surely something must change as you learn, but what?

With remarkable foresight, Cajal proposed a hypothesis. He suggested that when signals are transmitted across the gaps between neurons (via neurotransmitters, as was later discovered), the brain learns by changing the efficiency of that propagation. Nearly fifty years after Cajal's prediction, the Canadian psychologist Donald Hebb further developed this idea, proposing the following rule. (Fig. 1.15).

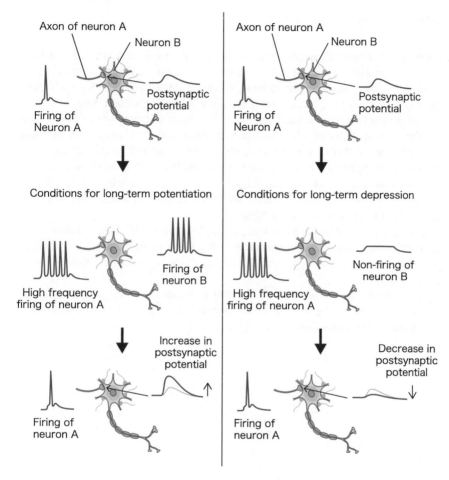

Fig. 1.15 Hebb's law

Hebb proposed that the magnitude of postsynaptic potential varies according to the input to and output from a neuron. There are two cases. When neuron A fires, and neuron B fires along with it, the postsynaptic potential increases (Fig. 1.15). Conversely, when neuron A fires but B does not, the postsynaptic potential decreases. At its heart, the rule is very straightforward: When input from neuron A contributes to the firing of neuron B, future firings of A will more easily result in B firing. If A's firing does not result in B's firing, however, future input from A will be less likely to cause B to fire. More succinctly, "neurons that fire together wire together." The rule was later named "Hebb's Rule."

Hebb's Rule was experimentally demonstrated in 1966 by the Norwegian scientist Terje Lømo, who showed that postsynaptic potential in rabbit brains strictly follows Hebb's predictions.

You Are Nothing But a Pack of Neurons

I hope that this chapter has given you a good idea of how neurons work. The description involved many technical terms that may be hard to take in all at once, but the main point here is not the specific biological mechanisms. Rather, I wanted to show that the neurons in your brain employ no magic tricks to establish "you."

The greatest mystery in neuroscience, and the central theme of this book, is that despite the simplicity of what individual neurons do, when present in enormous numbers, the result is an unpredictable phenomenon: the self. In this sense, Crick's "you are nothing but a pack of neurons" has a double meaning. The first is just what it says on the surface: at the most basic level, we are remarkably simple. But the phrase also conveys a sense of awe for the human brain—that a bundle of something as simple as neurons can produce "you."

How does the self arise from such a simple mechanism? This question lies at the core of the science of consciousness. When Crick eagerly entered this field, his predecessors gave him a warning: don't expect this to be as easy as determining the structure of DNA.

So where do we stand now, a little over a decade since the passing of Crick? In the next chapter, we will look into conventional experimental approaches to understanding consciousness.

Chapter 2
Pursuing the Shadows of Consciousness

Neuronal Activity Accompanying Consciousness

In the previous chapter, I described how subjective experience, or qualia, forms the consciousness that science addresses. Here, we go one step closer to the true challenges of understanding consciousness by examining how experimental neuroscientists have studied it in recent decades. The key concept is correlation between neural activity and consciousness, which applies a straightforward experimental logic: given that neural activities support consciousness, then changes in those activities should accompany changes in the content of consciousness. This approach was introduced to mainstream neuroscience by Nikos Logothetis, the founding father of experimental approach to consciousness.

Logothetis directed the Department of Physiology of Cognitive Processes at the Max Planck Institute for Biological Cybernetics in Tübingen, Germany. As acknowledged by his peers, he is a brilliant researcher who has produced remarkable experimental results. I have had the pleasure to work in his group for over a decade, in an attempt to follow in his footsteps in the field of visual consciousness.

So what does it take to extract neuronal activities correlated with visual consciousness? Even when animals are under anesthesia, their neurons respond to visual stimuli. In fact, most experiments were performed under anesthesia in the early days of visual neuroscience.

There is, thus, a need for some method of discerning neuronal activities that are associated with consciousness from those that occur even under anesthesia. This is where bistable visual stimuli come into play. As the name suggests, these are stimuli that cause perception to switch over time. As an example, stare for a while at the left shape in Fig. 2.1, which is called a Necker cube. Every few seconds, your perception as to which face is in the forefront should switch. Something similar happens when you view Rubin's vase on the right; the foreground and background alternate over time, so that the image sometimes appears to be a vase and sometimes two faces facing

© Springer Nature Switzerland AG 2022
M. Watanabe, *From Biological to Artificial Consciousness*, The Frontiers Collection,
https://doi.org/10.1007/978-3-030-91138-6_2

Fig. 2.1 Images inducing
bistable perceptions: (left) a
Necker cube, and (right)
Rubin's vase

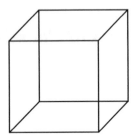 

each other. It is important to note that the two percepts do not simultaneously occur—
you cannot see a face of the cube as being both its front and its back, nor can you
see both the faces and the vase at the same time.

Bistable visual stimuli are highly valued because they induce transitions in the
content of conscious vision, whereas the external stimulus itself remains constant.
Because the stimulus does not change, visual processing unrelated to consciousness
should remain constant. Conversely, if we find brain activity that modulates along
with bistable perceptions, these activities are likely related to consciousness.

Binocular Rivalry

Among the many stimuli that induce bistable perceptions, the most powerful is
binocular rivalry, which I described in Chap. 1.

The predominant feature of binocular rivalry is that consciously perceived images
completely replace each other. In most other bistable stimuli, the perceptual switch
occurs in the secondary attributes, such as perception of depth (Necker cube)
or of foreground and background (Rubin's vase). By contrast, the whole image
switches in binocular rivalry. While you see one stimulus, the other completely
disappears. It provides an ideal condition for examining the neural mechanism of
visual consciousness.

In the many years I have studied consciousness, I have often been asked how
I can conduct research on something that is even ill-defined. When faced with
such questions, I bring up the example of binocular rivalry. There is no doubt
that the stimuli presented to each eye are input to the brain. Furthermore, as I will
describe in more detail below, both the visible and the invisible images trigger neural
activity even in the highest level visual areas. Under these conditions, what we
want to decipher are the requirements for such neural activity to rise to the level of
consciousness.

Logothetis's Search for Consciousness

In the mid-80's, Logothetis's bold plan was to record neural activity while monkeys were experiencing binocular rivalry. His experimental logic was straightforward: by exploring neural activity correlated with perceptual switches, he could isolate brain areas involved in visual consciousness. However, the experiment contained inherent risks. For one, there was no guarantee that monkeys would experience binocular rivalry. And even if they did, they would not necessarily be able to let us know it was occurring.

One of the barriers he faced was the fact that monkeys have short attention spans. In typical studies at the time, a single experimental trial consisting of stimulus presentation, perceptual report, and reward delivery lasted only a few seconds. If a monkey had to maintain concentration and wait for a reward too long, it would become annoyed and abort the trials.

When humans experience binocular rivalry, on average, the images alternate every two to three seconds. Assuming no large difference in timing between humans and monkeys, this implied that cycling through even a few bistable percepts would require ten seconds. No known studies had exposed monkeys to tasks lasting that long, so there was no assurance that they would ever manage. It should thus come as no surprise that many of his fellow scientists tried to dissuade him from this line of inquiry, for fear of what it might do to his career.

Logothetis ignored them, however, and started training the monkeys. The first step was to train them to maintain their gaze towards a dot on a screen, a necessity for awake behavioral studies that is closely related to "retinotopy," a neuronal property that I will describe in more detail later. In this task, monkeys were rewarded if they could fixate on the dot until it disappeared. Tasks were less than one second at first, then gradually became longer. Training for even this simple task may easily take months.

The next step was to train the monkeys to distinguish between two types of visual stimuli, using two levers to report what they saw (Fig. 2.2). They were trained to press the left lever when they saw one type of stimulus, and the right lever when they saw another. At this stage, both eyes were presented with the same stimuli, which were then swapped for a different one, thereby mimicking the perceptual alterations that occur during binocular rivalry. This intermediate training step is necessary because if stimuli for binocular rivalry were used from the beginning, it would be impossible to tell whether the monkeys were actually reporting their percepts, or they were just playing with the levers; the monkey only knows the timing of perceptual switches, if any, leaving no room for a proper rewarding strategy.

After further months of training, a monkey reached the point where it could continually report its percept for over ten seconds. Logothetis then changed the computer program to expose the monkey to binocular rivalry stimuli for the first time. And to meet his long-awaited expectation, the monkey began pressing levers in an unusual, anomalous rhythm. As it turned out, this odd rhythm would save his ambitious project.

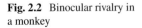

Fig. 2.2 Binocular rivalry in
a monkey

To meet the high standards of scientific experimentation, it was necessary to demonstrate that monkeys were indeed experiencing binocular rivalry. Even if they operated both levers to report a switch, this would not be a sufficient demonstration of binocular rivalry; it might be that they were only experiencing alternating mixtures of the two images. The experimental logic relied on the supposition that the invisible images were completely erased from consciousness, so if the monkeys were seeing some mixture of both, then this entire line of inquiry would be rendered moot.

However, Logothetis had prepared for this problem. During training, the monkeys were occasionally presented with an amalgam of both stimuli and trained to let go of both levers at such occurrences. Consequently, during exposure to binocular rivalry stimuli, monkeys only removed their hands from both levers for short periods of time between perceptual switches, thus assuring that the monkey only saw one stimulus or the other for the majority of time. Such transient mixed perception also occurs in human subjects (as you might have experienced when looking at Fig. 1.4), so it was even considered as further evidence that monkeys were indeed experiencing binocular rivalry.

An Upheaval in the Science of Consciousness

Even so, the world of science is a harsh one, and the deeper one treads into unexplored territory, the more one is exposed to the scrutiny of peers. As might be expected, some detractors remained unconvinced that Logothetis's experimental paradigm and results were sufficient to prove that monkeys experienced binocular rivalry. For example, the monkeys may have experienced some kind of unexpected mixing of the images that they were not exposed to during training. After all, they could not provide verbal reports like humans do.

Here is where that unusual, anomalous rhythm of lever presses came to the rescue. Logothetis was able to show that this unrhythmical rhythm was highly similar to what occurred in humans. Specifically, no fixed relationship existed between the duration of one perception and the next; whether a given perception lasted for a long or short time had no effect on how long the next one would be. There was one other similarity between monkeys and humans: while most perceptual durations were relatively short, occasionally a very long one occurred.

Considering these similarities, the chances were low that humans and monkeys were experiencing completely different visual phenomena, and thus, the peers could safely assume that monkeys indeed experienced binocular rivalry as we do. Now assured that the required experimental conditions were fulfilled, Logothetis needed only to insert electrodes into the monkeys' brains and, for the first time since the question was posed several thousand years ago by Greek philosophers, explore the brain mechanisms underlying consciousness. Indeed, this led to a string of achievements for Logothetis.

But before describing those findings, I would like to explain how the brain processes vision so that we can fully appreciate them. We again start with a pair of Nobel laureates.

Hubel and Wiesel

Thirty years before Logothetis's experiments, the Canadian scientist David Hubel and the Swede Torsten Wiesel had the good fortune to make one another's acquaintance. After receiving their Ph.D. degrees, both took positions in the laboratory of Stephen Kuffler at Johns Hopkins University. At the time, Kuffler had already produced many excellent studies on the retina, applying a method he developed to obtain electrical recordings from an intact eye. This allowed visual stimulation of the retina under the same conditions in which live animals see (Fig. 2.3).

Kuffler's discoveries were roughly as follows. There are two types of retinal ganglion cells (a type of neuron in the retina): an on-center type that emits action potentials when exposed to a bright region contained within a dark circle and an off-center type that emits action potentials by dark regions within a bright circle (Fig. 2.4).

Following Kuffler's achievements, Hubel and Wiesel recorded neural response in the primary visual cortex, which, as the name implies, is the location at which visual signals enter the cerebral cortex (Fig. 2.5). To perform their measurements, they inserted recording electrodes into the primary visual cortex of anesthetized cats. In accordance with Kuffler's findings in the retina, they first applied on-center and off-center stimuli. From there, Hubel's memoirs from his Nobel Prize lecture describe how they luckily arrived at their historical finding:

> Our first real discovery came about as a surprise. We had been doing experiments for about a month. We were still using the Talbot—Kuffler ophthalmoscope and were not getting very far; the cells simply would not respond to our spots and annuli. One day we made

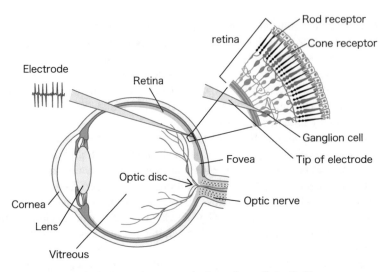

Fig. 2.3 Electrophysiological experiment on retinal ganglion cells by Kuffler

Fig. 2.4 Visual stimulus response characteristics of retinal ganglion cells

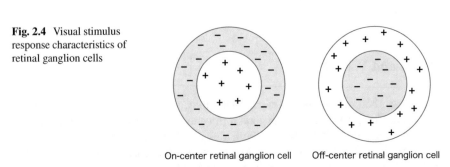

On-center retinal ganglion cell   Off-center retinal ganglion cell

an especially stable recording. (We had adapted my technique for recording, which used long-term implantations and the Davies closed chamber, to the short-term experiments, and no vibrations short of an earthquake were likely to dislodge things.) The cell in question lasted nine hours, and by the end we had a very different feeling about what the cortex might be doing. For three or four hours we got absolutely nowhere. Then gradually we began to elicit some vague and inconsistent responses by stimulating somewhere in the midperiphery of the retina. We were inserting the glass slide with its black spot into the slot of the ophthalmoscope when suddenly over the audiomonitor the cell went off like a machine gun. After some fussing and fiddling we found out what was happening. The response had nothing to do with the black dot. As the glass slide was inserted its edge was casting onto the retina a faint but sharp shadow, a straight dark line on a light background. That was what the cell wanted, and it wanted it, moreover, in just one narrow range of orientations.

The stories of great people often suggest that they are blessed with good fortune. Or perhaps, becoming great requires also luck. The latter cannot be discounted, considering the number of scientists who have come so close to great discoveries only to find them forever elusive. An example of this is the research group of Richard Jung, who had a laboratory in Freiburg, Germany. Already in 1952, seven years before

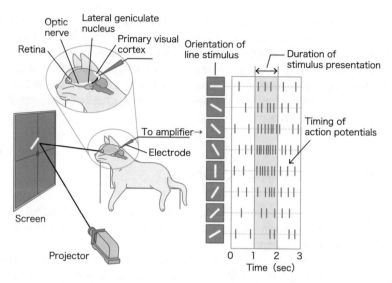

Fig. 2.5 Experiments on the cat primary visual cortex by Huber and Wiesel

Hubel and Wiesel's attempts, the group had successfully recorded neural activity in the cat primary visual cortex. They had furthermore spent years developing complex devices capable of presenting a variety of stimuli, including oriented line segments. Even so, despite conducting experiments for around a decade, they never stumbled across the correct answer.

According to Jung, this was partly because their intricate devices were inflexible. For example, they were incapable of making fine adjustments to the orientation of presented line segments. Yet, Hubel and Wiesel could present only one orientation, the fixed orientation of the slide edge, so it was sheer luck—that it matched the preference of the neuron they were recording that day—that led them to success in advance of Jung.

Coming back to science, through Hubel and Wiesel's discovery, we learned that while neurons in the retina respond to points, neurons in the primary visual cortex respond to line segments. On the face of it, this was a simple discovery, but it marked a new epoch in our understanding of cortical sensory processing.

Hubel and Wiesel initially used point-like stimulation because they believed that was appropriate: the scientific common sense of the time held that response characteristics in the retina should be carried over as-is to the primary visual cortex (V1). In actuality, however, hierarchical signal processing has already kicked in during the short trip from the retina to V1. Visual representations in the retina are somewhat similar to artwork by a pointillist painter, but as images make their way to the cerebral cortex, they are refined into fine line segments resembling a charcoal sketch. The core of Hubel and Wiesel's discovery was demonstrating that sensory representations in the brain evolve rapidly as they climb the processing hierarchy.

Hubel and Wiesel invited Kuffler to their laboratory to demonstrate their finding, which would later lead to their receiving a Nobel prize. While Kuffler reportedly found this discovery to be of some interest, he focused more on their omissions, such as not analyzing neural onset latency (the time between presentation of a stimulus and the resulting neural activity), claiming that was the sort of data that defined neuroscience. The episode vividly demonstrates how far ahead of their time Hubel and Wiesel's findings were.

Neurons Conduct Telephone Surveys

To understand how point-like information from the retina is transformed into line segments in the primary visual cortex, we need to recall the functional properties of neurons described in Chap. 1.

As explained there, information in the brain is transmitted as electric spikes, called "action potentials" (Fig. 1.6b). Neurons receive these action potentials, and their synapses tally up a positive or negative value. If the resulting sum exceeds a certain threshold, the neuron outputs its own action potential. Here, the sending side, from the viewpoint of a synaptic cleft is called a presynaptic neuron, and the receiving side is called a postsynaptic neuron.

This thresholding property of a neuron functions as something like a tele-phone survey. A postsynaptic neuron has dedicated phone lines connecting it to many presynaptic neurons, and receives input from them. Each input is determined according to whether the corresponding presynaptic neuron fired. Let's call the response a "1" if the presynaptic neuron fired, and a "0" if not. Inputs received from presynaptic neurons are weighted according to the amplitude of synaptic potentials arising in the postsynaptic neuron, and whether the grand total exceeds the threshold determines the final result of the survey.

So individual neurons are constantly conducting surveys by this simple mecha-nism, where the content of the survey depends on how the dedicated phone lines are distributed and how the inputs are weighted. The following two examples illustrate this concept.

Where Are the Cherry Trees Blooming?

When the cherry trees are blooming in Japan, television news shows maps tracking the blossoms as they open from south to north. Let's imagine exploring the cherry blossom front by performing a two-stage telephone survey.

In the first stage, we telephone a number of households (presynaptic neurons) within a given prefecture and ask whether the cherry trees have bloomed in their neighborhood (Fig. 2.6a). We score responses of "yes" as "1" and responses of "no" as "0." Multiplying each response by a weighting of $+1$, we can sum all the scores

Fig. 2.6 a A telephone-survey model of neuronal behavior, part 1: investigating cherry blossoms within a prefecture. **b** A telephone-survey model of neuron behavior, part 2: determining regions in which cherry trees have bloomed from the results of the survey in part 1

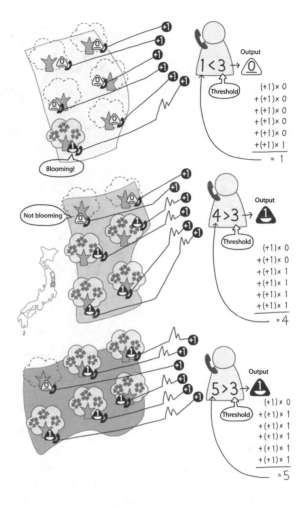

to get a picture of whether cherry trees are currently blooming in the prefecture. Furthermore, if we set the threshold for the surveyor (postsynaptic neuron) as half the number of households responding to the questionnaire ("3" in the figure), then the surveyor's output would represent whether at least half of the cherry trees in the prefecture have bloomed. The output is "1" if more than half have bloomed, and "0" otherwise.

In the second stage, we poll the surveyors in stage one (Fig. 2.6b) to determine in which region the cherry blossom front is situated. This stage is different in that we assign a weighting of +1 to responses from surveyors in southern prefectures of each region, and −1 to those in the north. Furthermore, surveyors conducting these regional surveys are assigned a threshold slightly less than half the number of prefectures in the region ("2" in the figure). Consequently, as in the example of Fig. 2.6b, only one region would output 1 and the others 0.

Fig. 2.6 (continued)

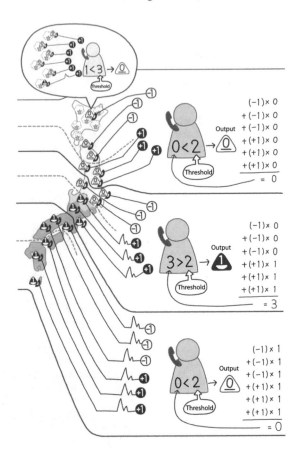

This second-stage telephone survey is somewhat peculiar, in that it relies on the fact that Japan is located in the northern hemisphere, so cherry trees bloom from south to north. If there are no cherry blossoms within a region, then the response from all of its prefectures will be 0, and so will be the weighted sum. Meanwhile, if cherry is blooming in all of its prefectures, then the negative values from the north would cancel out the positive values from the south, again giving a weighted sum of 0. The weighted sum becomes positive only when there are cherry blossoms in the south, but they have not yet advanced all the way north within the region

.

The Survey in the Brain

Information processing by neurons in the brain works in a similar manner, by laying out a network of dedicated "telephone lines." Fig. 2.7 is a diagram from Hubel

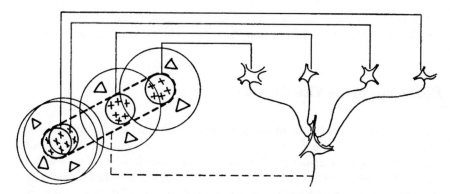

Fig. 2.7 Neural wiring from an on-center neuron to a line segment neuron. On the left side, the visual stimulus response characteristics (on-center type) of multiple on-center neurons are aligned with the outside world. On the right, line segment neurons react to external line segments by receiving positive synaptic input from connected on-center neurons (Reprinted from Hubel and Wiesel 1962)

and Wiesel's historical paper, where they explain how responses of line segment neurons can be formed. As the figure shows, a postsynaptic neuron that responds to line segments receives inputs from multiple presynaptic neurons that respond to on-center stimuli. Notably, the responding sites of these on-center neurons are linearly arranged within the visual field.

Assuming that these on-center neurons reside in the retina, they should be linearly arranged on the two-dimenstional retinal surface. This is because spatial relationships in the outside world are retained in the inverted image projected on the retina. When you look at, say, a house, its image passes through your eye's lens, and its inverted image is projected onto your retina, thus preserving the original topology (Fig. 2.8).

In actuality, however, things are not so simple. Neural projections to the primary visual cortex do not come directly from the retina. Rather, input arrives via part of a subcortical structure called the lateral geniculate nucleus (LGN) (Fig. 2.5).

This leads to an interesting question: Are horizontal and vertical relations retained in visual areas, just as they are in the retina? It certainly seems that retaining this topology would vastly simplify the neural wirings required for visual processing.

Fig. 2.8 Visible objects in the outside world are projected onto the retina upside down and backwards

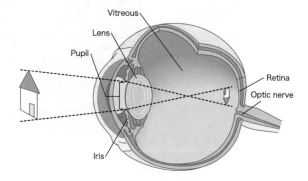

Visual features such as straight lines, curves and T-junctions are defined locally, so keeping visually adjacent information also adjacent in the cortex seems beneficial.

Retinotopy in the Visual Cortex

With apologies to the squeamish, I would like to describe an experiment that directly questioned whether retinal topologies are retained in the primary visual cortex. Researchers began by showing an image of concentric semicircles (Fig. 2.9 left) to a monkey for several hours. During this time, the monkey was anesthetized and unable to move its eyes, so its retinas received visual input free of spatial shifting. The monkey was then immediately euthanized and its brain extracted. The extracted brain underwent special treatment to stain only those neurons that were activated just before the monkey's euthanasia.

The right side of Fig. 2.9 shows the result of staining. Activated neurons form a semicircular pattern highly similar to that of the presented image, appearing as though the stimulus pattern was directly projected onto the cortex, leaving no doubt that topology is retained in the primary visual cortex. This property of maintaining spatial relationships with the retina is called retinotopy.

Then what about downstream visual areas? Fig. 2.10B shows that they too exhibit retinotopy, or to be more precise, the boundaries between distinct visual areas are defined using the fact that these downstream areas are indeed retinotopic.

Here, the cerebral cortex is only a few millimeters thick, and its wrinkly appearance comes from the cerebral cortex getting folded up in our body's effort to create as much surface area as possible within the cramped confines of our skulls. If a human cerebral cortex were spread out, it would be about the size of a newspaper page. Figure 2.10C shows a flattened human cerebral cortex where each of the differently shaded areas depicts a distinct visual area.

Once a one-to-one correspondence between the outside world and the retinal image has been established, preserving positional relations while passing images

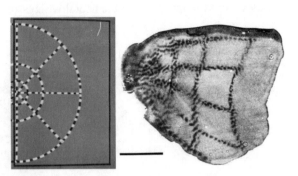

Fig. 2.9 Retinotopy in the monkey primary visual cortex. The line segment at the bottom indicates 1 cm (Adapted from Tootell et al. 1988, Copyright 1988 Society for Neuroscience)

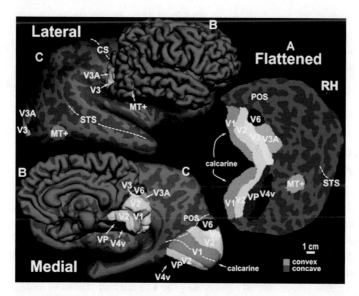

Fig. 2.10 The hierarchical structure of human visual brain areas. The two figures on the left are pairs of images showing the cerebral cortex in its original folded state (B) and spread out using dedicated software (C). The figure in the upper left shows the cerebrum from the outside, while the figure in the lower left is a view from the inside. The right side (A) shows a spread out, flattened view cut through the calcarine sulcus, which passes through the middle of the primary visual cortex (V1) (Adapted from Pitzalis et al. 2006, Copyright 2006 Society for Neuroscience)

on to downstream visual areas is not too difficult; all that's needed is straight and untangled wiring between the areas. Indeed, the axons connecting these areas are aptly aligned. These straight connections are in fact formed while we are still a fetus. Chemical substances guide axons as they grow, leading them as they stretch from one visual area to another. The characteristics of these substances slowly change along the cortical sheet in a two-dimensional manner, resulting in a straight and topology preserving connection.

Box 1: Why Retinotopy is Vital to Visual Processing

Strictly speaking, there would be no need for maintaining retinotopy if it were possible to make the neuronal connections arbitrarily complex. As illustrated by the telephone survey analogy, the fundamental nature of neural information processing is in the wiring between neurons. So long as the wiring is preserved, it does not matter where in the brain those neurons are situated.

In actuality, however, arbitrarily complex wiring is not possible because it requires more space. The cerebral cortex—the so-called "gray matter" in which the soma, axon terminals and dendrites of neurons are contained—is a surface layer just a few millimeters thick. The remaining "white matter" consists entirely of axonal fibers connecting various parts of the brain. Even with orderly straight wiring that maintains retinotopy and allows nearby neurons to share incoming wiring, the brain nearly overflows with connective axons.

Hierarchy of Visual Areas and Response Properties of Neurons

As Hubel and Wiesel demonstrated, on the way from the retina to the primary visual cortex, the preferred stimulus of neurons (the type of visual stimulus they respond to) transforms from points to line segments. But what happens from there?

Visual processing in the brain splits into two streams, the ventral ("underside") stream and the dorsal ("backside") stream (Fig. 2.11). Previous research has shown that the ventral stream mainly processes the shapes and colors of things we see, while the dorsal stream processes their position and movement. Here, we focus on the ventral stream and describe how visual processing progresses in the cerebral cortex. The dorsal stream will be highlighted in the next chapter.

The ventral stream begins at the primary visual cortex (V1), then arrives in turn at the secondary (V2), tertiary (V3), and fourth (V4) visual areas and ends at the infer-otemporal (IT) cortex. Neuronal responses gradually increase in complexity in the following manner: V2 continues to respond to straight line segments in fundamen-tally the same manner as V1. One notable difference, however, is that neurons here strongly respond to illusory edges (called "subjective contours") like those we saw in the neon color spreading demonstration (Fig. 1.3) in Chap. 1. Complexity increases in V3 and V4, where neurons respond to features such as angles, curves, and inter-secting lines. When visual information finally arrives at IT, we find neurons that respond only to specific objects, such as faces and hands. These are the exceptions,

Fig. 2.11 Stimulus response in the dorsal and ventral streams of the monkey visual cortex

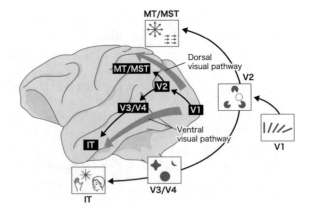

Fig. 2.12 The visual
alphabet (Tanaka 1996,
Reproduced with permission
from the Annual Review of
Neuroscience, Volume 19 ©
1996 by Annual Reviews,
http://www.annualreview
s.org)

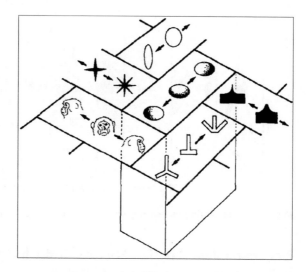

however; we now know that most neurons respond to moderately complex shapes like those shown in Fig. 2.12. Such properties were discovered by Riken researchers Keiji Tanaka and Kang Cheng, along with Ichiro Fujita at Osaka University, who named these moderately complex forms "visual alphabets."

IT neurons break down visual stimuli into a combination of component shapes, just as we can break down words into alphabets. This makes the handling of myriad visual stimuli much easier, reducing the number of neurons needed to represent the visual world.

In fact, the initial discovery of neurons dedicated to faces and hands was met with great surprise. After all, if the brain were to prepare neurons for every object we might see, there would never be enough. Such a strategy would furthermore require allocating new neurons every time we saw something new, and the time required to do so would surely place tight restrictions on the brain's ability to cope with the ever-changing environment. Meanwhile, visual alphabets do not run into such problems, where novel objects only require novel combinations of component shapes.

Finally, I briefly describe how the preferred stimulus of neurons can become increasingly complex along the visual hierarchy. This is not particularly difficult, similar to how points become lines as visual stimuli travel from the eye to the primary visual cortex. As Fig. 2.13 shows, all that's needed are dedicated neuronal connections, allowing upstream visual features to spatially combine and characterize more complex downstream features.

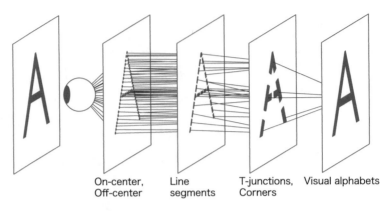

On-center,   Line   T-junctions,   Visual alphabets
Off-center   segments   Corners

Fig. 2.13 Neural wiring for realizing increased complexity of visual response

Generalization: Another Hierarchical Change in the Response Properties of Neurons

As the response properties of neurons become increasingly complex along the visual hierarchy, there is another development: responses become more generalized.

Simply put, neuronal responses get "sloppier." For example, line-segment preferring neurons discovered by Hubel and Wiesel in the primary visual cortex also exist in the secondary visual cortex, but are less sensitive to small changes; they maintain heightened activity in a wider range of line orientations. The degree of generalization increases as we climb the visual hierarchy, and it is particularly pronounced in the IT region, the highest-level visual area in the ventral stream. Face neurons in macaque IT are activated regardless of whether they are viewing a photograph or an illustration, a human face or a monkey's. Facial orientation is also generalized, with responses occurring regardless of whether faces are looking straight forward or off to the side. As you might have guessed, response generalization in IT is not limited to face and hand neurons—it also applies to visual alphabets, where neural activity is robust to the deformation of moderately complex visual components.

Moreover, generalization also occurs with respect to retinotopy. Individual neurons fire only when stimuli appear within a particular region in the visual field, the so-called "receptive field." Generalization of retinotopy occurs in the form of receptive fields becoming increasingly large with each successive visual area (Fig. 2.14). Here, the size of a receptive field is measured in visual angles, where one degree is equivalent to a one-centimeter-long object held approximately 56 cm from the eye. In the macaque visual system, the size of the receptive field is just a fraction of a degree in the primary visual cortex, but increases as we proceed to higher levels, reaching tens of degrees in IT.

Interestingly, such increases in receptive field size also occur within a single visual area. In fact, generalization of neuronal activity was first reported by Hubel and Wiesel in the primary visual cortex. Figure 2.15 shows another illustration from

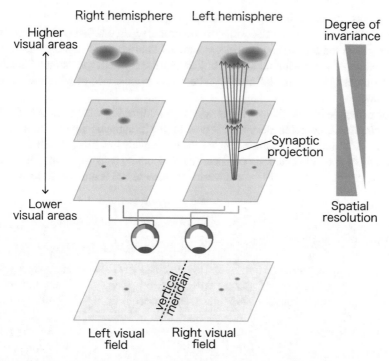

Fig. 2.14 Hierarchical structure of visual areas and an overview of neural wiring. The circular regions in each visual area depict regions that receive input from point regions on the retinal surface. These circular regions become increasingly large along the visual hierarchy, which results in the increase of receptive field size

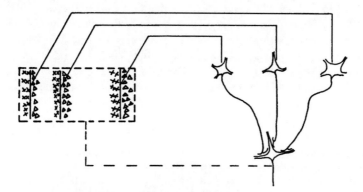

Fig. 2.15 Within V1 retinopic generalization realized through neural wiring from simple to complex neurons. (Reprinted from Hubel and Wiesel 1962, Copyright 1962 Society for Neuroscience)

their historic paper explaining how retinotopic generalization may occur within the primary visual cortex. They discovered two neuronal types, one that is highly sensitive to changes in the positions of line segments and another that "sloppily" allows for some extent of the positional shift. They called the former type "simple" neurons and the latter type "complex" and hypothesized the existence of within-V1 wiring as follows.

The principle behind this wiring is very simple, forming what is called an "OR" logic circuit. A complex neuron receives multiple synaptic inputs from simple neurons. Assuming that complex neurons have relatively low firing thresholds, input from any of the connected simple neurons will cause it to fire. This OR circuit principle can be applied to the generalization of other visual features described above.

Box 2: Neuronal Response Characteristics Are Formed Through Postnatal Learning

As mentioned above, the cortical wiring that produces retinotopy is innate (formed before birth). It is thus interesting to consider whether neuronal wiring requiring higher precision, such as those between "point" and "line segment" neurons, are also innate, or whether they are acquired (formed after birth) through learning.

Colin Blakemore and Graham Cooper performed experiments on kittens in an attempt to answer this question. They first raised kittens in complete darkness from two weeks after birth, then for five months placed them in an environment that consisted only of vertical stripes, five hours per day (Fig. 2.16, left). The kittens even wore collars that prevented them from seeing their own bodies. At other times they were returned to pitch darkness.

The right side of Fig. 2.16 illustrates neuronal properties of the primary visual cortex after five months of the above treatment. The angles of line segments represent the preferred orientations of recorded neurons. While many neurons responded to vertical lines, there were basically none that responded to horizontal ones. And as expected, when a stick was placed horizontally in

Fig. 2.16 A cat raised in an environment of only vertical stripes (Adapted from Blakemore and Cooper 1970)

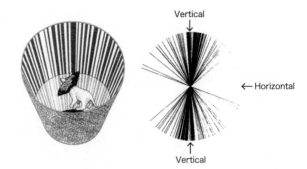

front of these cats, they would stumble on it, as if they could not see them at all.

These experiments demonstrate that preference for line segments in the primary visual cortex is indeed acquired through visual experience. Certainly, one might consider it to be quite straightforward; there are limits to the extent to which chemical substances can guide axons during development.

Another interesting fact was that this dramatic effect occurred only in kittens subjected to the experiment immediately after they were born; almost no such effect was observed in kittens who began the treatment at even just a few months old. The significance of these findings extends far beyond the scope of visual processing. They may explain, for example, why only young children can acquire language just by listening to it, and why true mastery of, say, playing the piano requires training from a very young age. Year by year, we learn more about the molecular mechanisms by which learning occurs in the brain and the inner workings of such "critical periods" are not an exception. Someday, fundamental research may even lead to new genetic treatments that allow child-like enhanced learning in adults.

Searching for the Seat of Consciousness

Now that we have laid out the background, let's proceed with Logothetis's binocular rivalry experiments.

Having established that monkeys experience binocular rivalry, the recording part is rather straightforward. Simply insert an electrode in the area of interest while the monkeys report their percepts. Although there is one thing that needs to be taken care of; the stimulus presented to one eye needs to match the recorded neuron's preference, whereas the stimulus presented to the other eye should *not*.

Once this requirement is met, analysis is again straightforward. We just need to focus on the neuron's preferred stimulus and sort out periods when it entered the monkey's conscious vision and when it did not. If there is a significant difference in firing rates between the two, we may conclude that the recorded neuron is associated with conscious vision. Figure 2.17b shows an example of just such a neuron. The vertical axis shows the neuronal firing rate (number of firings per second), which varies perfectly in sync with the perceptual reports indicated in the bar below.

Note that this is just a single example. The important question is what proportion of the neurons is modulated within a given area. Logothetis and his colleagues thus spent years making hundreds of measurements from various visual areas in monkeys.

Logothetis's first target was the highest visual area in the ventral stream, IT. There, over 80% of neurons were highly modulated with perceptual switches (Fig. 2.17c). One might then wonder if IT could be the much sought-after seat of consciousness,

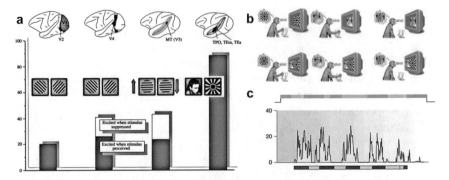

Fig. 2.17 Neuronal responses in monkeys experiencing binocular rivalry. **a** Perceptual report during training (upper row) and while experiencing binocular rivalry (lower row). The rightmost figures in the upper and lower rows, respectively, show a physically mixed image and one that is perceptually mixed during binocular rivalry. In both cases, the monkey releases both levers (Adapted from Logothetis 1998). **b** An IT neuron. Remarkably, its firing rate changes according to the monkey's perceptual report (Adapted from Blake and Logothetis 2002). **c** Proportion of neurons linked to bistable perceptions by visual area (Adapted from Logothetis 1998)

the simplest definition of which being the location where neural information perfectly agrees with conscious vision. If there is a perfect agreement, then one could expect some conversion mechanism from neural activity to consciousness, thus explaining consciousness with neither excess nor deficiency. (But wait until chapter four, where we discuss the "hard problem" of consciousness!)

IT is unlikely to fulfill this definition, however. Recall that the remaining ten-plus percent of neurons continued to fire at the same rate, irrespective of which stimulus entered conscious vision. Another critical point is that even during times when stimuli were not entering consciousness, a majority of IT neurons maintained increased activity as compared to a separate experimental condition where monkeys faced a blank screen. So while IT is the highest level in the ventral stream, its neuronal firing overall does not coincide with conscious vision; conscious and subconscious processing appear to coexist. Thus, the dividing line between consciousness and subconsciousness cannot be drawn at the spatial scale of cortical areas.

Box 3: Do "Elite Neurons" Exist That Constantly Take Charge of Consciousness?

To approach this question, we must first ask whether any fixed division exists between consciousness and subconsciousness, or whether their separation is dynamic and alters depending on context.

David Leopold, one of Logothetis's disciples and currently a researcher at the National Institutes of Health, and Alexander Maier, Leopold's student at the Max Planck Institute, conducted an ambitious research to answer this question.

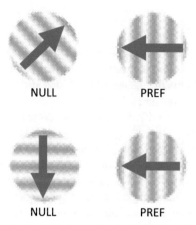

Fig. 2.18 Two pairs of visual stimuli used by Leopold et al. during binocular rivalry experiments. Two types of stimulus were prepared for the left eye (left, top and bottom), while the right eye was presented with a stimulus in the preferred direction of the recorded neuron (the direction that maximizes the neuron's firing rate; right, top and bottom) (Adapted from Maier et al. 2007, Copyright (2007) National Academy of Sciences, U.S.A.)

Their focus was a visual area called the medial temporal (MT) area (Fig. 2.11), a high-order dorsal stream area that responds to visual motion. Logothetis and his colleagues, in a previous research project, had recorded MT neurons, and found that about half modulated their activity during binocular rivalry. What Leopold and Maier wanted to know was whether that half was fixed. If so, this would imply the existence of "elite" neurons that are always responsible for a given content of visual consciousness.

To test this, Leopold and Maier decided to induce exactly the same contents of visual consciousness under two stimuli configurations. Specifically, the stimulus presented to one eye was a moving grating shared between the two configurations that matched the preferred direction of the recorded neuron, whereas to the other eye, different stimuli were presented for the two configurations (Fig. 2.18). This clever scheme allowed the researchers to evoke a single common visual experience during the times when the shared moving grating entered consciousness, but with a different overall stimulus configuration.

Measurements performed under these conditions showed that even when the contents of consciousness were identical, many neurons that were perceptually modulated in one stimulus configuration (e.g., the upper pair in Fig. 2.18) were not modulated in the other (e.g., the lower pair). These results proved that the neurons responsible for a given content in visual consciousness were context dependent, and thus, the neuronal level boundary between consciousness and subconsciousness is even fluctuating within a visual area.

The Battle of V1

Experimental evidence of consciousness and subconsciousness coexisting in the highest-level visual area raises an intriguing question: Can this coexistence be traced backward, all the way to the first stages of the visual stream? Or are some low-level visual areas in the brain dedicated solely to subconscious visual processing, with no room for consciousness at all? The primary visual cortex became the testing ground for these two conjectures.

Whether the primary visual cortex, the entry point of visual information in the cerebral cortex, plays a part in conscious vision will profoundly affect the way we view the neural mechanism of consciousness. If it turns out that it does play a role, the chances are that consciousness, in general, is spread across all regions of the cerebral cortex. If it does not, then the front lines of visual consciousness must be located in further recesses of the visual system, and consciousness would be something that can be cornered within the cortical hierarchy.

By the way, there is another reason why for decades now the primary visual cortex has been a primary battleground for the study of visual consciousness: a paper by Francis Crick and Christof Koch published in Nature in 1995. In that paper, Crick and Koch made the bold prediction that the primary visual cortex does not support consciousness, an idea that many scientists strongly resisted.

They presented two reasons for this hypothesis: first, the fact that there are no direct neural projections to the prefrontal cortex, and second, that the primary visual cortex contains much information that never rises to the level of consciousness.

Regarding their first reason, Crick and Koch imagined, at the time, the existence of a cortical region that is ultimately responsible for consciousness, a role they ascribed to a region called the prefrontal cortex. The prefrontal cortex is a high-level brain region that performs decision-making, situated at an even higher level than the highest visual areas of the ventral and dorsal streams, bridging them, together with other sensory systems, to the motor system. Assuming the prefrontal cortex as the seat of consciousness, they postulated that visual areas related to consciousness should send direct neural projections to it to "present" it the visual world.

Put briefly, there are two problems with this line of thinking. First, if there is some "homunculus" in the brain ultimately responsible for establishing consciousness (here, the prefrontal cortex), then there must be another homunculus within the first that is responsible for establishing its consciousness, and so on, leading to an infinite regress. Second, there have been many reports of patients with lesions to their prefrontal cortex without associated loss of subjective visual experience. Koch later revised his thinking, publicly recognizing that the prefrontal cortex is not critical for establishing conscious vision.

Information in the Primary Visual Cortex that Consciousness Cannot Access

The more fundamental issue, and one that remains hotly debated today, is their second argument for why the primary visual cortex does not support consciousness: the fact that much information contained there does not rise to the level of consciousness.

Plenty of such information does indeed exist, as remarkably shown in the optical illusion on the left side of Fig. 2.19. Hard though it may be to believe, the squares marked A and B are exactly the same shade of gray, as demonstrated by connecting them with gray bars. The brightness of the two squares appears different because your brain is trying to show you the "true" light reflection properties of them, which requires compensating for the influence of the shadow cast by the cylinder. Your brain surmises that if B is reflecting light at the same brightness as A despite being in shadow, then B must have a larger surface reflectance, and it is this reflectance that rises to the level of consciousness.

Your brain's attempts to show an object's inherent light reflectance properties by eliminating factors such as differences in lighting or the presence or absence of casted shadow is called "color constancy," and this effect only shows up in the fourth visual cortex and beyond. It thus follows that A and B—which appear so very different to our mind—evoke equivalent neural activity at earlier stages of the visual pathway. This in turn demonstrates that these areas, including the primary visual cortex, contain raw brightness information that our consciousness cannot access.

Here is another example: Our eyes are rapidly moving even when we gaze at a fixed point, and the amount of movement is nonnegligible. This phenomenon is called microsaccades, and if our conscious vision directly reflected these movements, we would be experiencing something like a shaking seen in a hand-held camera. In fact, high-speed measurements of eye movements suggest that your field of view would shake so badly that you would find it impossible to read this book. However, we remain completely unaware of these movements, even if we try.

In contrast, neural activity in the primary visual cortex is fully subject to the effects of microsaccades, rising and falling with each fine flicker of the eye. Hence, we must assume that some sort of motion correction mechanism exists downstream of the

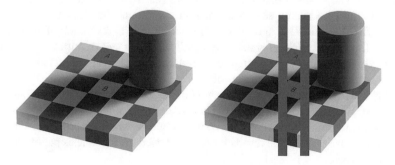

Fig. 2.19 An optical illusion demonstrating color constancy (© 1995 Edward H. Adelson)

primary visual cortex, and only the corrected visual information rises to the level of consciousness. Again, the raw uncorrected information present in the primary visual cortex fails to enter conscious vision.

Phenomena such as those described above leave little room for doubt that the primary visual cortex contains information that we cannot consciously access, and indeed this is widely accepted. The controversy regarding the assertions by Crick and Koch arose because they went one step further, claiming that no information in the primary visual cortex could be consciously accessed.

Binocular Rivalry in the Primary Visual Cortex

In 1996, after Crick and Koch made their claim, Leopold and Logothetis went on to publish a paper on binocular rivalry in the primary visual cortex. They found that around one-tenth of neurons were significantly modulated with perceptual switching during binocular rivalry (Fig. 2.17c, but note that the results shown there also include those from the secondary visual cortex). Even that one-tenth showed only minimal modulation, detectable only after averaging the results from tens of perceptual switches. Put another way, the majority of neurons in the primary visual cortex faithfully responded to physical stimuli, not varying their activity in accordance with the monkey's conscious vision. Visual information represented there was nothing like the all-or-nothing switching of perception.

But still, in regard to Crick and Koch's claims, the ten percent of neurons that Logothetis found responding to perceptual switches may take on a new meaning; possibly, that one-tenth—and that one-tenth only—had the potential to rise to the level of consciousness. Indeed, when Leopold and Logothetis published their results of V1, they focused on that aspect and titled their paper "Activity changes in early visual cortex reflect monkeys' percepts during binocular rivalry."

Later experiments made the position of Crick and Koch even more untenable, particularly when Frank Tong and his colleagues used functional magnetic resonance imaging (fMRI; see Box 4) to measure neural activity in humans experiencing binocular rivalry.

We obviously cannot insert electrodes into the brains of healthy humans; we need a noninvasive technique to make measurements, one that does not damage the brain or the skull. fMRI is one such method with a spatial resolution of around one cubic millimeter. Cortical gray matter generally contains tens of millions of neurons crammed into that space, and signals captured by fMRI are considered to be the sum of overall neural activity within the cube.

Tong's fMRI measurements of human V1 during binocular rivalry showed approximately 100% modulation of activity. That is, the signal doubled in strength during periods when stimuli entered subjects' conscious vision, as compared to periods when stimuli were invisible. These results further weakened the hypothesis put forward by Crick and Koch.

As an aside, why did the results of Logothetis and those of Tong differ so drastically regarding the magnitude of modulation? Why did Logothetis's results indicate that only 10% of neurons changed their activity even slightly, while Tong's showed an overall doubling of activity?

There are two plausible reasons for this difference. The first is possible differences between how consciousness relates to the primary visual cortex in monkeys versus humans. The second is differences in measurement methods; Logothetis was recording individual neurons while Tong used fMRI.

Taking an interest in this problem, Maier and Leopold performed measurements recording both single neurons and fMRI on monkeys experiencing binocular rivalry. The results showed that differences between the two results were due to their measurement methods: Maier and Leopold's neural recordings agreed with those by Logothetis, while their fMRI measurements agreed with those by Tong. Thus, we must conclude that different measurement methods were capturing different aspects of neural activity.

Neural recordings using electrodes were unequivocally capturing the action potentials that neurons emit—in other words, their output. So what was fMRI measurements capturing? At the time, scientists already knew that fMRI signals do not reflect changes in firing rates of neurons, but rather their synaptic inputs (see Box 4). Taking note of this fact, the combined results of all three groups suggested that whether or not a stimulus entered conscious vision had only little effect on the firing of neurons in the primary visual cortex, though somehow their synaptic inputs greatly changed. A strange set of results, indeed, but my own research, which I will describe next, provides a possible interpretation.

Box 4: How fMRI Works

The principles of fMRI were discovered by Seiji Ogawa when he was working as a researcher at Bell Labs. Considering the scientific contributions from fMRI and its potential for future clinical applications, his discovery certainly deserves a Nobel Prize, and many Japanese neuroscientists, myself included, eagerly await that good news every year. Together, Nikos Logothetis, the star of this chapter, later made significant contributions toward decoding fMRI signals. Below, I trace the footsteps of these two scientists to understand the principles of fMRI and the origins of its signals.

In 1988, Ogawa was researching ways to use magnetic resonance devices to produce images of biological tissue. One day, he noticed some black specks on a brain cross-section he had imaged, and after looking into them, realized that they were sections of blood vessels. Furthermore, he found that oxygen concentrations in those blood vessels determined the darkness of the spots. This discovery transformed magnetic resonance from a specialized tool for anatomical imaging in biological and medical fields to one that could be used to measure neuronal activity, "functional MRI."

Meanwhile, it had been known since the early twentieth century that blood-flow levels vary with neural activity. In primitive early experiments, scientists placed stethoscopes on the heads of subjects who were solving difficult math problems, and heard increased blood flow. Later developments such as positron emission tomography (or PET, a method that uses radioactive isotopes to measure cerebral blood flow) showed that such blood flow increases are localized in the immediate vicinity of neuronal activity. Due to this property of brains, expectations for fMRI to push the limits of non-invasive brain measurements were high. Indeed, after Ogawa's invention of fMRI, its contribution to cognitive neuroscience increased year by year.

This is where Logothetis comes in. His studies of binocular rivalry in monkeys had resulted in scientific acclaim worldwide, not to mention many job offers. He found himself struggling to decide whether to join one of the Max Planck Institute in Tübingen, Germany, or the Massachusetts Institute of Technology in Boston. In the end, he chose Germany, drawn by guarantees of a large research budget through his retirement at age 67 (later extended to 72 in honor of his outstanding scientific contribution) and acceptance of his audacious research plan: developing a monkey-specific fMRI device that would allow simultaneous neuronal measurements using recording electrodes.

The foremost advantage of fMRI is its noninvasive nature, so many scientists were mystified as to why someone would want to use it on monkeys. Regarding the origin of fMRI signals, the general assumption at the time was that fMRI signal strength was correlated to the firing rates of neurons. Some had even proposed conversion formulas, such as equating a 5% stronger fMRI signal with a 30 Hz increase in neuronal firing rates.

Meanwhile, even with extensive institutional support, the development of the monkey fMRI device Logothetis envisioned was no simple matter. While the device itself was fully developed by 1997, measuring neural activity in the device remained difficult. fMRI captures blood oxygen concentrations using a huge fluctuating magnetic field in the order of Tesla, but neuronal measurements using electrodes must capture electrical action potentials of just a few tens of microvolts. Simultaneously performing both measurements meant finding a way to suppress the extensive signal artifacts caused on the latter by the former.

A dedicated hardware development team of over ten members, including Axel Oeltermann who later became the final author of their highly cited Nature paper, finally achieved this goal after grueling years of work. The results were highly surprising. As it turned out, the signals that fMRI captures are due less to the output of the neurons themselves and more to the input they received through synapses.

Logothetis's research called attention to the mechanisms of neural activity and increased blood flow, establishing " neurovascular coupling" as a major field in neuroscience. Currently, there is a promising hypothesis that glial cells (which surround neurons but are not neurons themselves, because they do not produce output) receiving synaptic input from neurons trigger muscles surrounding fine arteries to adjust the blood flow rate.

The Difference Between Correlation and Causality

Before describing my own work and its underlying motivations, I need to provide some background. In a broad sense, evidence from binocular rivalry experiments suggest that neural activity in various visual areas, including the primary visual cortex, is correlated with conscious vision. Can we therefore conclude that these areas causally relate to consciousness? Generally speaking, in scientific interpretation of experimental data, we are not allowed to equate correlation with causality.

To give a familiar example, summertime electric power usage correlates with ice cream sales: as one increases, so does the other. That does not, however, mean that one is causing the other. Of course, demands for the electricity needed to produce and store ice cream will increase power consumption to some extent, but only by trivial amounts in the context of overall power usage. Much more important is the shared cause, the outdoor temperature. As it gets hotter, more people use air conditioning, which drives up electricity consumption. At the same time, they eat more ice cream to cool down.

Indeed, finding ways to distinguish between correlation and causation is an issue common to all of science, and neuroscience is not an exception. Moreover, the ultimate goal of the science of consciousness is to pinpoint the neural mechanisms that give rise to consciousness, and the first step toward doing so is identifying causality. Searching for mere correlations between neural activity and consciousness is only the first leg of the journey. Crick and Koch foresaw this problem in the 1990s, and introduced "neural correlates of consciousness" (NCC) as an initial research target for the scientific investigation of consciousness. As we will see below, despite the appearance of written words, its definition dives deep into causality.

NCC and Akira

Some readers might be familiar with *Akira*, a Japanese animated film from the 1980s. In that movie, a boy named Akira with powerful psychic powers is blown apart, but parts of his brain are artificially kept alive in a hideous contraption comprising countless tubes. Interestingly enough, the brain fragment alone, as long as it entered a particular dynamic state, was sufficient to deploy its dreadful psychic power.

When I first learned about Crick and Koch's definition of NCC, this movie sprang to mind. Specifically, they defined NCC as "the minimal set of neuronal events and mechanisms jointly sufficient for a specific conscious percept." Note how this "minimal and sufficient" constraint goes beyond the realm of correlation.

NCC is like Akira's brain fragments, which were able to express their awesome powers from both within his head and within the device they were later contained in (Fig. 2.20). The dynamics of the brain fragment was fundamental, not the vessel or the boy's body, which served only to contain it; the brain fragment could deploy its psychic power regardless of whether it was supported by the boy's heart, lungs and the surrounding brain neural circuitry, or supported by oxygen pumping machinery and electronics, so long as it was able to maintain a particular state. NCC is similar. So long as NCC, a particular dynamic state of a neuronal network, is established, so too will be subjective experience Note here that, as we will further see in the next chapter, the definition of NCC goes beyond mere causality, in the direction of "criticality" .

To clarify the above, let's consider the subjective experience of an apple placed in front of you. Is the retina included in this definition of NCC? After all, if you close your eyes, not only the apple but the entirety of your visual world disappears from your consciousness. When you once again open your eyes, the apple reappears in your visual field and its subjective experience is reestablished. Retinal activity exhibits a perfect correlation with subjective experience. Moreover, the retina is playing a causal role in generating subjective experience. Considering only these matters while we are awake, the retina, which can be thought of as an extension of the brain, seems essential for the formation of subjective visual experience, and thus, included in NCC.

However, you can also dream about apples, and when you do so, your external environment is isolated from your brain. Dream apples require no aid from your eyes; your brain fully fabricates them for you. It is replaced by some neural mechanism that provides alternative input to the NCC. For this reason, our retina can be safely excluded from the NCC.

Pursuing NCC by Process of Elimination

Importantly, NCC does not necessarily comprise modulated neural activities during binocular rivalry as in the above example of our retina during wakefulness. So how can we close in on NCC? One hint lies in dreams, which served as a condition to prove that the retina is not part of NCC.

As Fig. 2.21 shows, anything included in NCC should be correlated with changes in consciousness under all conditions. In contrast, for anything not included in NCC, conditions may exist under which no such correlation is seen (in the case of the retina, these conditions are the dream state). Perhaps, then, we can test new conditions one after another, excluding non-NCC until hopefully only the NCC remains. In other words, we might derive NCC through a process of elimination. To close this chapter,

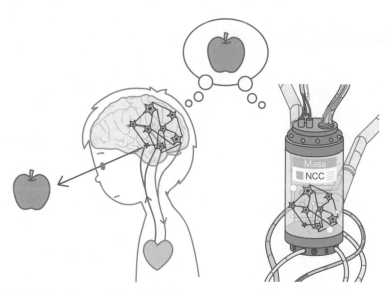

Fig. 2.20 NCC in a brain and NCC in a jar

Fig. 2.21 The quest for
NCC through a process of
elimination

neural activity in correlation
with conscious percept
under condition A

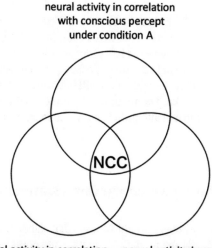

neural activity in correlation
with conscious percept
under condition B

neural activity in correlation
with conscious percept
under condition C

I describe a human fMRI experiments that I performed with my colleagues at RIKEN
and Max Planck Institute in Tübingen, following just such an approach.

We aimed to use "visual attention" as a factor that provides the desired new
condition. Try reading the horizontal and vertical lines in Fig. 2.22 while keeping
your gaze on the dot in the middle. You will probably notice that, as you attempt

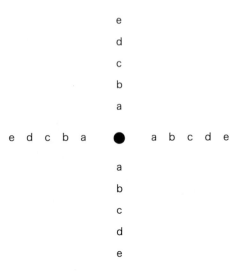

Fig. 2.22 Visual attention freely moves around the gaze point

to read a certain character, you focus your processing capacity on that character. This is visual attention. It allows for the selection of some information and inhibition of others, which leads to increased speed and accuracy of processing for the selected targets.

In daily life, the target of gaze and target of visual attention often match, but as you have just experienced, we can shift our spotlight of attention away from our line of sight. In fact, we even have four spotlights of attention.

We chose visual attention as a factor for a reason. Namely, if we do not control for visual attention, then attention will be naturally directed at stimuli entering conscious vision, thus risking conflation of its effects with those of consciousness. Indeed, the presence of visual attention is known to increase neural activity in the primary visual cortex.

CFS: An Innovation in Consciousness Studies

What made things difficult for us is that standard binocular rivalry is not suitable for independently manipulating attention and consciousness. This is because its bistable nature requires subjects to report their percepts; otherwise, it would be impossible to tell which stimuli were entering conscious vision at which times. And if subjects report their percepts, they cannot avoid directing attention to the target stimulus.

Simply put, we needed an experimental configuration that renders stimuli invisible, as in binocular rivalry, but without the perceptual switching.

Fortunately for us, and for many other scientists in the field, in 2005, Koch and Naotsugu Tsuchiya developed a method called continuous flash suppression (CFS) (Fig. 2.23). In that method, different stimuli are presented to the right and left eyes, as in binocular rivalry, but a special configuration prevents perceptual switching from

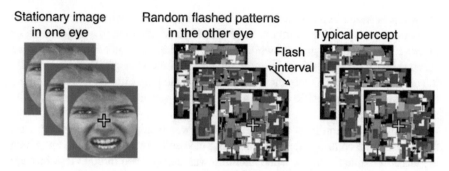

Fig. 2.23 Continuous flash suppression (Adapted from Tsuchiya and Koch 2005)

occurring. Subjects thus remain conscious of one of the visual stimuli, but never see the other.

This occurs through asymmetry in the intensity at which stimuli are presented to each eye. One eye is presented with a low contrast static image (Fig. 2.23, left), while the other is presented with images comprising over a hundred colorful rectangles whose positions vary at a rate of up to ten times per second (Fig. 2.23, right). These rectangle images somewhat resemble artwork by Piet Mondrian, so Tsuchiya, the inventor of CFS, named them "Mondrians." As a result of this strong asymmetry, subjects consciously and continuously perceive the more intense Mondrians, but never the less intense static image.

I still vividly remember the day when Tsuchiya, then a Ph.D. student of Christof Koch, came rushing into the kitchen of Shimojo lab, where Ryota Kanai (a Shimojo lab member who will appear in Chap. 5) and I were conversing over coffee. Tsuchiya called us into his office to demonstrate the soon-to-be-well-known CFS effect. Since Koch lab and Shimojo lab both occupied the same basement floor of the Beckman Institute situated at the west end of the Caltech campus, it was only a thirty-second walk. But during that short walk, we sensed Tsuchiya's sheer excitement, and could not help but expect something big.

After entering his office, Tsuchiya told us to make sure that we keep both eyes open while looking into his fancy dichoptic viewing setup (similar to that in Fig. 2.2, but built for humans). After telling him we could see only colorful rectangles flashing at high speed, he told us to close one eye. Upon doing so, we finally saw the face of a furious man staring at us! We immediately foresaw its impact on the field, and congratulated Tsuchiya for his great invention. Indeed, as of early 2022, Tsuchiya and Koch's CFS paper has nearly 1000 citations.

Consciousness Versus Attention

Having resolved the issue of entanglement between perceptual reporting and visual attention, we next needed a method for manipulating visual attention. We decided to switch between behavioral tasks performed by subjects, which is the standard method for manipulating attention. To guide subjects' attention to a target stimulus (a low-contrast moving grating) while maintaining their gaze on a fixation point, we asked subjects to report the visibility of the target stimulus by pressing a button. On the other hand, to guide their attention away from the target stimulus, we instructed them to search for a specific letter in a rapidly changing character string used as the above fixation point.

In summary, visibility of the target stimulus was manipulated by applying CFS (together with a control condition consisting of identical visual material, but which the target remained visible), while attention was manipulated through task instruction. Combinations of these allowed four stimulus–task conditions. With the above configuration in place, we could determine whether responses in the primary visual cortex were related to attention, to consciousness, or to both.

Empire Strikes Back

Finally coming to the results, our observation revealed that neural activity in the primary visual cortex is not modulated by visual consciousness, only by visual attention. This is depicted in the right panel of Fig. 2.24, which shows the results from a subject under four conditions of visual consciousness and visual attention regarding the target stimulus: with–with, with–without, without–with, and without–without. For all subjects, we observed no difference in fMRI signal for visibility of target stimuli, whereas the signal basically doubled when attention was directed towards the target. Applying the elimination strategy described earlier, we may therefore exclude the primary visual cortex from NCC.

I wish I could tell you that we reached the experimental conclusion under a carefully worked-out plan as described, but it was more of a thorny path where the initial results were not so straightforward.

In truth, when we started the project, we did not pay any attention to attention. Our initial motivation was simply to investigate the effects of visual consciousness in the primary visual cortex under conditions of no perceptual report. The results, however, were puzzling: we found no trace of visual consciousness in the primary visual cortex as reported in dozens of previous papers. The fMRI signal remained the same, regardless of whether the target stimulus entered subjects' conscious vision. These were novel results, but with a missing piece. After all, no journal will accept a paper that says, "This is what happened under our experimental conditions, but we don't know why things came out this way." Thankfully, visual attention saved the day. By additionally controlling attention, we were able to reproduce the doubling of fMRI signals of Tong et al.

As an aside, it's a shame that modern science pressures researchers to "shut up and publish." This attitude diminishes romantic aspects of science, such as the exchange of personal communications among rival groups about half-baked ideas and results, as was common until around the mid-twentieth century. It conceals the drama behind scientific creation, which can be full of insight. Today, researchers instead simply compile the results of their experiments, wring out the most appropriate research proposition, and compile it all into a dry report. The resulting paper belies all the messiness involved in its evolution, which could be very beneficial to scientific progress if shared among research groups.

That aside, let's attempt to interpret the results from a higher perspective. Of course, this is just a single study, and we should not over-generalize its results, but it certainly suggests that conscious vision does not involve the lowest visual area in the cerebral cortex, whereas visual attention does. In this regard, it is quite reasonable for visual attention to tap into the earliest possible visual system, since its goal is to increase the speed and accuracy of visual processing. Indeed, it has been shown that even LGN, the subcortical structure that takes in retinal input and relays it to the primary visual cortex, is involved in visual attention and modulates its activity accordingly. Meanwhile, it is sensible to assume that the neural mechanisms of visual consciousness evolved to block the primary visual cortex altogether, as it is packed with too much "raw" visual information. We will address this matter in Chap. 5, after I introduce a general hypothesis on consciousness.

Finally, regarding the mystery of discrepancy between the results of neuronal recordings and fMRI measurements, it all makes sense if the previously observed modulation during binocular rivalry stemmed from visual attention. Visual attention, where the spotlight of attention is manipulated by our minds, is evidently a top–down effect. So it makes sense that V1 fMRI signals are more affected by changes in attention, since it better reflects the top–down synaptic input from higher-level visual areas. To explain the discrepancy, the spiking of V1 neurons would not be modulated by attention as much, as demonstrated in studies that recorded neurons while monkeys performed a visual attention task.

As a final note, Tsuchiya contacted us on the day that our paper was published. He was in Washington D.C. at an academic conference, he said, raising a glass of

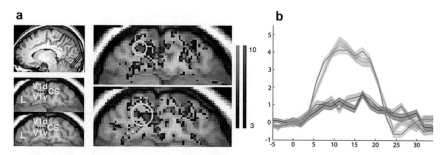

Fig. 2.24 FMRI brain activity in the human primary visual cortex under independent manipulation of visual consciousness and visual attention. **a** The darkened area in the circle is the area in the primary visual cortex responsible for the target stimuli. **b** While somewhat difficult to read, this graph shows only the brain activity in the above brain region under conditions of attention with and without consciousness (upper two lines) and of no attention with and without consciousness (lower two lines) (Adapted from Watanabe, Cheng, et al. 2011)

champagne with Koch. No doubt Koch was overjoyed to hear our results, which supported the position he and Crick had suffered so much criticism for over the years. My only regret is that I could not see Crick's response; he had unfortunately passed away in 2004.

I later attempted to collaborate with Koch, as I will describe in the following chapter. When I first visited his office in Seattle, I noticed a large photograph of Crick on the wall. It was signed, "I will be looking after you."

Chapter 3
Neural Manipulation for Causal Investigation of Consciousness

Transcranial Magnetic Stimulation as a Revolutionary Tool

The latter half of Chap. 2 introduced the neural correlate of consciousness (NCC), a concept proposed by Crick and Koch as an initial research target for experimental approach to consciousness. So how can we further close in on NCC? Since its very definition deeply involves causality, we may turn to experimental manipulation, a scientific tool used to address causality.

Taking the previous summertime electric power usage and ice cream sales as an example, how could we tell whether the observed correlation is due to a shared cause or there is indeed a causal relationship between the two? For instance, if the local government ordered the closing of all ice cream stores for a day, but still the electric power usage modulated just the same, we may conclude that, at any rate, the ice cream sales was not causing the change in power usage. Of course, such social manipulation experiments are unrealistic, but this is exactly how causality is addressed in modern neuroscience. Scientists artificially alter neural activity and investigate the resulting effects on other neural activity, behavior and such, in an attempt to reveal causal relations between the manipulated activity and brain functions. Regarding NCC, manipulation experiments enable us to go beyond what we saw in the previous chapter, merely correlating neural activity and perception. In this chapter, I discuss the past and future of investigating NCC through manipulation experiments, starting with the development of TMS, a noninvasive method for manipulating neural activity that has greatly impacted psychology.

TMS is an acronym for "transcranial magnetic stimulation." As the name implies, it entails the use of a powerful electromagnet to apply a dynamic magnetic field from outside the cranium. Abrupt changes in the magnetic field create electric currents in the brain that directly stimulate neurons. Experimental psychologists have long relied exclusively on sensory stimuli and perceptual reports in their attempts to unravel the mechanisms of the mind, but as TMS was introduced in the mid-1990s to mainstream psychology, it became a revolutionary tool that expanded experimental freedom.

© Springer Nature Switzerland AG 2022
M. Watanabe, *From Biological to Artificial Consciousness*, The Frontiers Collection,
https://doi.org/10.1007/978-3-030-91138-6_3

Fig. 3.1 TMS devices by Thompson (1910; left) and Barker (1985; right) (Reprinted from Scholarpedia)

Despite the rather short history of applying TMS in scientific research, its developmental history is surprisingly long. Researchers such as Jacques-Arsène d'Arsonval and Silvanus Thompson had already created prototypes by the late nineteenth century. The photo on the left in Fig. 3.1 shows Thompson's device, in which subjects placed their head between two massive electromagnets. Those brave enough to do so were subjected to a sudden release of the device's built-up electric current, and reportedly experienced vague visual hallucinations. Despite its size, however, the device was unable to generate a magnetic field strong enough to trigger neuronal firings in the brain, and today scientists believe the hallucinations were due to direct stimulation of the retina. It was only much later, in 1985, that Anthony Barker produced the first device capable of inducing cortical neuronal activity (Fig. 3.1, right).

I was first exposed to TMS while spending a sabbatical year in the laboratory of Shinsuke Shimojo at the California Institute of Technology. One day soon after I had arrived, when I was still living in a motel on the outskirts of Pasadena, a doctoral student named Daw-An Wu invited me into a dark room. In one corner hulked a sinister piece of electronic equipment, connected by a thick cable to a magnetic coil shaped like a figure eight. No sooner had I realized that this was the TMS device I had heard rumors about, than Wu said, "Let's start with motor functions," and placed the coil on top of my head. He flipped a switch, the machine began to whine, and the voltage level displayed on a LED indicator started to climb. When the charge indicator finally lit up, Wu gleefully punched the button. The machine issued a startling bang, bang! and my arms flew up above my shoulders.

It is hard to express how jolting this was. I was expecting something more like a finger twitch. Apparently, I wasn't the only one. "Oops," Wu said, without a trace of regret, "I forgot we have to lower the voltage for Asians."

Next, he turned the lights off and placed the coil on the back of my head, right over my primary visual cortex. When he pressed the button, I saw white flashes (called "phosphenes") flying in the darkness. Although I already knew at an intellectual level that consciousness was nothing but a product of electrical activity in our brains, this was the moment that it truly sank into me.

The Yin and Yang of TMS

My sabbatical at Shimojo's laboratory was in 2003. Four years earlier, this same lab had produced a significant TMS study that established the reputation of Yukiyasu Kamitani, who went on to greatly expand the potential of fMRI measurements. He and Shimojo reported on TMS-induced illusory perceptions that were the opposite of those described above. In their experiments, TMS was applied to the primary visual cortex while subjects viewed a lattice pattern briefly displayed on a monitor. When TMS was synchronized with the presentation of this pattern, subjects perceived an illusory hole in the lattice, as shown in Fig. 3.2. Shimojo and Kamitani named this effect "TMS-induced scotoma."

"Scotoma" is a medical term describing partial loss of the field of vision due to brain injury. In the early twentieth century, the speed of bullets fired from rifles increased to the point where they could pass completely through a human skull. Perversely, this meant that more soldiers survived being shot in the head, but in most cases there were lasting side effects. When the bullet passed through cortical visual areas, the result was often partial blindness, coined scotoma.

During the Russo-Japanese War, a Japanese optometrist named Tatsuji Inoue reported compelling findings related to scotoma. Using his newly developed device for precisely estimating the positions of brain lesions based on bullet entrance and exit points, he found a systematic correspondence between locations of lesions and positions of partial vision loss as reported by patients (Fig. 3.3). This was actually

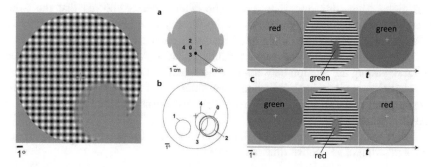

Fig. 3.2 TMS-induced scotoma. The left image shows a reproduction of what subjects perceived. The central image shows correspondences between stimulated sites and positioning of scotoma in the field of view. The right image shows that the TMS-induced scotoma is filled in with the background from the future (Adapted from Kamitani and Shimojo 1999)

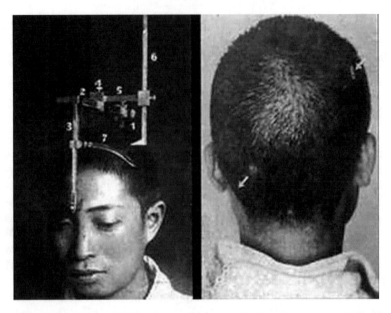

Fig. 3.3 Measurement of gun wound traces by Inoue Tatsuji (Reprinted from Inoue Tatsuji 1909)

the first reported evidence of retinotopy, described in the previous chapter, and it made Inoue the first internationally recognized Japanese neuroscientist.

Kamitani and Shimojo used TMS to noninvasively reproduce the bullet-induced scotoma that Inoue had studied. In doing so, they confirmed that changing the site of TMS stimulation altered the position of scotoma in the visual field (Fig. 3.2, center). This result alone would have made a fine paper, but the pair went on to determine just what filled the illusory hole; whether the filler came from the past or the future.

To investigate this, they presented subjects red screens before the lattice pattern and green ones after, and vice versa. The results clearly showed that the color presented after the lattice pattern filled the illusory hole (Fig. 3.2, right). That is, TMS-induced scotoma was filled in with sensory contents from the future.

To fully comprehend these peculiar results, I need to explain what we have learned so far regarding the relation between consciousness and time. This time–consciousness relation is closely related to an experiment that I describe at the end of this chapter, and it is also a vital element of the consciousness mechanism I discuss in the fifth chapter.

I start with two experiments by Benjamin Libet that fundamentally changed our understanding of subjective time and free will.

Time Lag in Consciousness

As described in Chap. 1, even a single iteration of neural processing consists of multiple stages of molecular mechanisms and requires time. As a result, it takes over a tenth of a second for the first volley of action potentials evoked by visual stimuli to reach higher visual areas, so our conscious vision should have a lag of at least that duration. Measurements by Libet, however, showed that the actual lag is far longer (Fig. 3.4).

Libet applied electrical currents to electrodes inserted into the cerebral cortex of patients going through open-brain surgery under local anesthetic. When he stimulated brain areas corresponding to skin sensations, subjects felt like something was touching their arm. Libet's initial goal was to determine the conditions necessary for

Fig. 3.4 Stimulation of the brain and hand by Libet

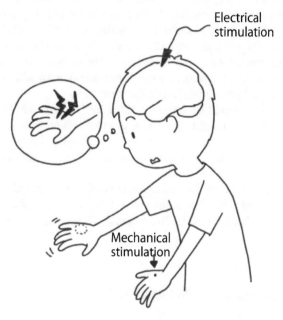

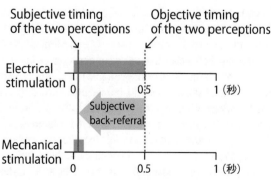

electrical stimulation to invoke these sensations and asked subjects to report their percepts as he varied the strength and duration of electrical stimulation. As a result, he discovered something perplexing: at moderate levels of electrical stimulation, skin sensations did not occur unless stimuli continued for over half a second.

This is unusual because it suggests that illusory skin sensations only arise half a second after the onset of electric stimulation, according to the following rationale. First, subjects themselves have no way of knowing how long any given stimulus will last. Second, they perceive nothing if the stimulus lasts less than half a second, implying that by the time they do feel something, the stimulus must have already been present for at least half a second. This is quite a long time in the context of neural processing, but one could possibly chalk it up to the special conditions of electrical stimuli being directly applied to the brain.

The real mystery lies in Libet's next experiment, in which he mechanically stimulated subjects' skin along with the electric stimulation, and asked them which sensation they perceived first. In the example shown in Fig. 3.4, peripheral skin stimulation is being applied to the left hand, while electrical stimulation is causing skin sensations in the right hand by stimulating the left cortical hemisphere. Since we already know that perceptions due to electrical stimulations are delayed by half a second, we might predict that the subjects should sense the skin stimulation first. But surprisingly, they reported perceiving both at the same time.

How should we interpret this? Libet's interpretation was that, just as with direct electrical stimuli to the brain, external stimuli sensed by peripheral receptors also take half a second to be consciously perceived. In plain English, he suggested that we are subjectively experiencing the world at a half-second lag behind the physical world.

Furthermore, Libet conjectured that when perceptions do arise, we feel as if they occurred at the time of their underlying stimulus. Libet referred to this point as "subjective back-referral." To understand its necessity, we must step outside of "passive" frameworks.

Subjective Back-Referral

A concrete example should make the necessity of "subjective back-referral," say, an internal time-rewinding mechanism, clear. To begin with, if we had no way of acting upon the external world, if we only experienced it passively, we would never notice such a lag between physical time and conscious time. It would be like watching a soccer match on television; even live television broadcasts suffer from a transmission delay of at least half a second, but we never notice it since we have nothing to reference the broadcasted events to. Things are quite different, however, when we can actively engage with the environment. Our self-initiated actions serve as a temporal reference for sensory input, and therein lies the necessity of such time rewinding mechanisms.

Let's look at the matter with actual numbers, using baseball as an example. Assume that a professional baseball player pitches at 85 miles per hour, so the ball arrives at the batter just 0.5 s after it is released. Say the batter swings, and the bat hits the ball. Under these conditions, who feels what, and when?

The situation for someone watching passively from the stands is unproblematic. This fan never notices the 0.5 s perceptual delay—the pitcher's pitch and the batter's swing are all viewed externally and passively.

That's not the case from the batter's perspective. It was the batter who made the decision to swing, and of course, he must have made this decision before the ball passed over home plate, specifically within the 0.5 s since the pitcher threw it. Similarly, he must have started to swing the bat within that 0.5 s, since there is no way he could have hit the ball if he had started swinging just as it passed him. Including the time needed for initiation of muscle movements and the bat to travel through the air, it seems reasonable to assume he started the swing 0.3 s after the pitch. Of course, the decision to swing must have been made even earlier.

So what is the batter perceiving when he hits the ball? At the instant when the ball and bat come into contact, the batter's vision, hearing, touch, and other physical sensations all work to convey the impact. However, based on Libet's findings, all of these senses are perceived 0.5 s after they have occurred. Adding things up, the impact must have been perceived at least 1.0 s after the moment of pitch (0.5 s perceptual delay + 0.5 s interval from pitch to ball–bat impact).

The problem here is the large time difference between the time at which the batter initiates his swing (0.3 s after the moment of pitch), which will serve as the above-mentioned temporal reference, and the time at which he perceives hitting the ball (1.0 s after the moment of pitch). If there were no internal time-rewinding mechanism that compensates for this time difference, the batter would feel as if it took a whopping 0.7 s from initiation of his swing to the multi-modal sensory experience of bat–ball impact. In physical reality, by this time, he would have long finished hitting the ball and already be on his way to first base.

Of course, actual perceptions follow a smooth narrative. The batter starts his swing, the bat curves around from above his shoulder, and he feels the impact of the ball hitting the bat, all with no odd lagging.

So Libet, out of necessity, introduced subjective back-referral to resolve the contradiction between the perceptual lag he had discovered and the continuous sensation of time regarding one's actions and perception: that when perceptions occur, we feel as if we experienced them at the time of their underlying stimulus.

Does the Future Affect Consciousness?

Conscious time has one other interesting feature: while lagging behind actual time, it is affected by what is the future of conscious time. This is no doubt confusing when put in words, so it is best to experience it yourself, along with a partner, through what is called the cutaneous rabbit illusion.

Following the A, B, C labels in Fig. 3.5, close your eyes and have your partner rapidly tap out either an A-A-B or an A-A-C pattern on your forearm. During this

Fig. 3.5 The cutaneous
rabbit illusion

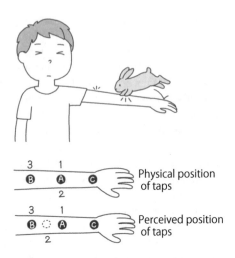

time, try to focus on where you feel the second tap. If you do this, you should find that if A-A-B was tapped out, you feel the second tap somewhere between A and B, while if the pattern was A-A-C you feel the second tap between A and C. In both cases, the perceived position of the second tap shifts toward the third.

If perceptual time lags were negligibly short and we feel the taps as soon as they occurred, displacement of perceived location should not happen. At the time when the second tap occurs, the third tap would still be in the future, so there would be no way for the second perceived position to shift towards the randomly chosen third position. Instead, the very fact that the second tap is perceptually relocated in the direction of the third unequivocally demonstrates that, indeed our conscious time is lagging behind physical time by a considerable margin—one so long that you perceive the second tap only after the physical occurrence of the third tap.

This perceptual misplacement occurs when the time interval between taps is from 0.04 to 0.2 s. While the upper limit of 0.2 s is quite less than the 0.5 s derived from Libet's experiments, it nonetheless suggests that the "now" of consciousness lags substantially behind the actual world. And coming back to our initial point, from the viewpoint of the conscious now, the future affects our sensory perceptions, as evidenced by the fact that the third tap affects the perceived second tap backward in time. So, conscious perception not only lags behind physical time, but it also extends cloud-like into the consciousness future, so that events that have not yet risen to the level of consciousness affect events that are currently occurring in our conscious mind.

Finally, returning to where we left off, namely Kamitani and Shimojo's experiments, the above view of conscious present expanding into conscious future explains why subjects perceived TMS-induced scotomas filled with the background color of the future.

What About Conscious Free Will?

If our subjective experiences do indeed lag 0.5 s behind the physical world, some might consider this to pose an existential risk to what matters most to many of us. Namely, what happens to conscious free will?

Recall the baseball example, in which the batter hits a pitched ball at 85 miles per hour. Recall also that it only took the ball approximately 0.5 s to reach the batter. This implies that the bat hit the ball as soon as it left the pitcher's hand in conscious time. If that is the case, then who judged the quality of the throw and decided to take the swing? It certainly seems that the batter's conscious self could not have done these things. There just isn't enough time! Therefore, Libet had to consider whether our conscious selves can possess free will.

To test this, he placed on test subjects' heads a device that could measure electric signals emitted from their brains (brainwaves). He furthermore attached to their wrists an electromyography (EMG) device capable of measuring even the slightest muscle movement. He then placed in front of them a device resembling a clock with a single rapidly spinning hand and instructed subjects to simply flex their wrists at any moment they wished, but only to note the position of the clock hand when they decided to do so. Figure 3.6 shows the setup for this experiment, along with a graph depicting the timing of the decision for wrist movement, the timing of actual wrist movement, and the time course of the recorded brainwave.

The first thing to notice is the delay of around 0.2 s between the decision to flex and the onset of wrist movement. This is not particularly surprising, since there are multiple stages of neural processing from the subject's internal decision to actual twist of the wrist. First, the subject's decision-making area, situated at the highest levels of the cortical hierarchy, triggers a motor command, which is sent to downstream motor areas and broken down into fine sequences for numerous muscle movements. Then the converted command is sent to the wrist muscles through the spinal cord and peripheral nerves, where the flex action is finally manifested. These processes require a significant amount of time. The controversial point here is the timeline over which the brainwaves increase; there is a noticeable ramping up of brainwaves starting 0.3 s before the decision to flex. In other words, the subject's brain had already made the decision to move the wrist well before the subject became consciously aware of the decision. A straightforward interpretation of this result is to state that no, we do not possess conscious free will.

Yet, in an effort to somehow retain the concept of free will, Libet proposed a notion he called "free won't." Even if our subconscious minds secretly plot to make decisions, Libet suggested, we retain something like a last-minute veto power by which we can override those decisions. Upon closer consideration, however, deciding whether to issue such a veto would itself require a similar preparation-and-decision period, which would likely take place somewhere inaccessible to the conscious mind.

Recent debate over this issue has tended toward abandoning pained attempts to cling to conscious free will, and instead accepting that it is likely an illusion.

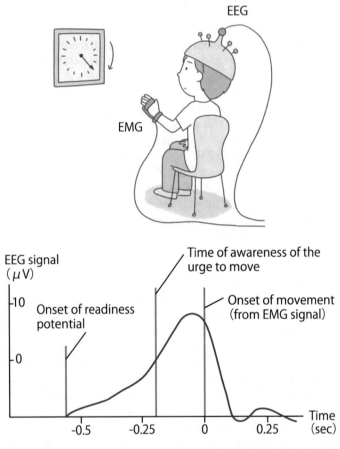

Fig. 3.6 Libet's free will experiment (lower graph adapted from Libet's "Mind Time")

Accordingly, attempts to rescue conscious free will are often criticized as summoning "ghosts from the past."

To see what this means, consider once again the participant undergoing Libet's experiment in Fig. 3.6. Somewhere in the brain, there should be some trigger that initiates wrist movement. This trigger is most likely some minimal number of neurons, which only slightly change their firing rate. It is natural to think that this small trigger gradually grows larger to the point where it enters the subject's consciousness and finally culminates in a signal for wrist movement. This all makes sense if we assume that the readiness potential reflects the build-up of this initial subconscious trigger.

The critical point here is the almost certain impossibility of being conscious of this initial trigger, that minimal number of neurons that only slightly change their firing rate. Under these conditions, it is quite challenging to assert conscious free will. Such assertions imply that consciousness refers to something that resides outside the brain, but can nevertheless freely alter brain activities, triggering the initial neural trigger in some ghostly manner.

This separate treatment of mind and body, which includes the brain, is called "mind–body dualism." Dualists assert that the true nature of consciousness is separate from the brain, in some cases even separate from the material world, yet can somehow communicate with it. Such dualistic theories are the "ghosts from the past" that I mentioned earlier.

René Descartes, whom I discussed at the start of Chap. 1, was a dualist. He suggested that the pineal gland mediates communication between the brain and consciousness, which exists somewhere outside the physical world (Fig. 3.7). He chose the pineal gland because while most brain structures have left and right components, there is only one pineal gland right at the center. He reasoned that since there is only one consciousness, there should be only one mediator. While this sounds nonsensical today, we can't really criticize Descartes. After all, neurons had not even been discovered in his day; the prevailing theory was that the brain controlled the body through cerebrospinal fluid with some hydraulic actuation mechanism. Supposedly, since Descartes was way ahead of his time and understood the true difficulty of relating consciousness to any physical matter including the brain (we will elaborate on this more in the next chapter), he was driven into a corner to believe such matters to be true.

Fig. 3.7 Descartes' mind–body dualism and the pineal gland, from his *The Passions of the Soul*

Conscious Free Will is a Grand Illusion

I may not have yet convinced you that there is no such thing as conscious free will. After all, it is a nearly unshakable aspect of human nature to feel that we ourselves make conscious decisions. Despite such convictions, however, some very interesting psychological experiments do indeed suggest that free will is an illusion.

Peter Johansson, Lars Hall and their colleagues conducted one such experiment (Fig. 3.8). Subjects were shown photograph pairs of female faces and asked to quickly indicate which they found more attractive. Subjects were then handed their chosen photograph and asked to explain the reason for their decision. On the surface, this seems like a simple preferences survey, but there was a trick: a few of the photographs were rigged so that it was replaced with the non-chosen face when it was returned to the subject. Intriguingly, most subjects never noticed the replacement, and went on to explain why they "chose" the face that they had not in fact chosen. Typical reasons given were "She looks like an aunt of mine I think, and she seems nicer than the other one," "Just a nice shape of the face, and the chin," and "I thought she had more personality, in a way. She was the most appealing to me." Every feature of these oral reports was the same as those for trials in which the photographs were not replaced,

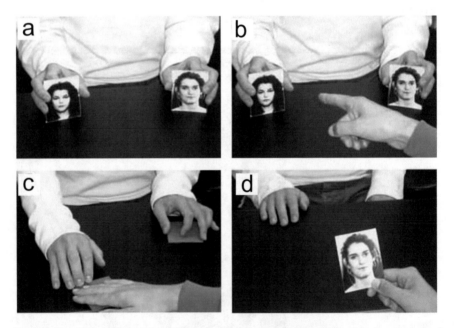

Fig. 3.8 The choice blindness experiment by Johansson and Hall. **a** Subject is shown a pair of facial photographs. **b** Subject points at the face they feel is more attractive. **c** Using sleight-of-hand, the tester passes the subject the photograph they did not select. **d** Subject describes why he "chose" this photograph (Reprinted from Johansson et al. 2005)

including factors such as verb tense usage, frequency of vague explanations, and length of description.

The results of this experiment suggest that the association between the neural mechanisms for decision-making and those for describing the reasons for the made decision is quite tenuous. It almost seems as if two parties are involved: one who makes subconscious decisions, and another, the one we are conscious of, who tries to invent post hoc reasons for why those decisions were made.

Johansson and Hall named this effect "choice blindness." This psychological phenomenon tells us that conscious free will is likely a grand illusion—a deception created by our brains.

Motion Phosphenes and NCC

Returning to the use of TMS in manipulation experiments, allow me to introduce two more TMS researchers: Alvaro Pascual-Leone and Vincent Walsh.

When TMS stimulates the primary visual cortex, the result—as I learned first-hand—is hallucinatory white flashes (phosphenes). However, Pascual-Leone and Walsh stimulated the MT/MST area, a high-level area in the dorsal stream (Fig. 3.9, top) that processes visual motion. The outcome is a different kind of hallucination called "motion phosphenes."

Injury to the MT/MST area can result in the full loss of motion perception: the world is perceived as a series of static images, like dancers beneath a strobe light. In TMS-induced motion phosphenes, fine moving patterns accompany the hallucinatory brightness (Fig. 3.9, bottom).

So what can we conclude from the fact that subjective experience is triggered by direct stimulation of high-level visual areas? Normally, we experience visual contents after a long chain of neuronal firings, starting in the retina and progressing through the primary visual cortex, then through intermediate and finally, higher-level visual areas. Yet we can skip all that and experience vision by directly applying TMS to the higher-level. In this case, none of those skipped areas—from the eyes up to pre-MT/MST—seem to be contributing to consciousness, so we might be tempted to conclude that these areas are not included in NCC.

Interestingly, things aren't quite so simple. One curious feature of neural wiring in the brain is that there are no one-way streets. Namely, signals never flow only from lower-level visual areas to higher ones; there is also feedback of visual information from higher-level areas to lower ones.

When higher-level areas are stimulated, such top–down flows of visual information can cause increased activity in intermediate- and lower-level areas. This is exactly the phenomenon that Pascual-Leone and Walsh wished to investigate. Specifically, they questioned whether top-down neural feedback to the primary visual area is necessary for conscious vision.

They focused on the primary visual cortex, conducting a curious experiment aimed at determining whether TMS induced scotoma (illusory holes in visual images) would

Fig. 3.9 Two types of phosphenes (phantom flashes) resulting from TMS stimulation. When TMS stimulation is applied to the primary visual cortex, the location of stimulation (A–H, top) determines where the phosphene appears (A–H, bottom). In contrast, stimulating the MT/MST area produces motion phosphenes, whose position is not much affected by the site of stimulation. The bottom figure shows the location of phosphenes and motion phosphenes in the field of vision; their opposite positioning reflects the fact that opposite brain hemispheres are stimulated (Adapted from Cowey 2005)

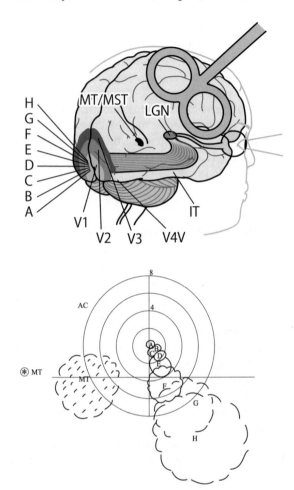

also appear in MT/MST triggered motion phosphenes, a sort of illusion within an illusion. If the perception of motion phosphenes required top-down neural activity in the primary visual cortex, then they expected to see TMS-induced scotoma as in Shimojo and Kamitani's experiment. Conversely, if activity in the primary visual cortex was unnecessary, then no illusory holes should be seen within the motion phosphene, no matter what manipulations they performed in the primary visual cortex.

Specifically, they stimulated subjects' primary visual cortex and the MT area at various timings, asking subjects to evaluate the intensity of the resulting motion phosphenes on a four-point scale ("I see nothing," "I see spots, but they are motionless," "I see spots, but I'm not sure if they are moving," or "I see spots and they are definitely moving.").

Figure 3.10 shows the results. The perceptual intensity of motion phosphenes was lowest under conditions where TMS stimulation of the primary visual cortex

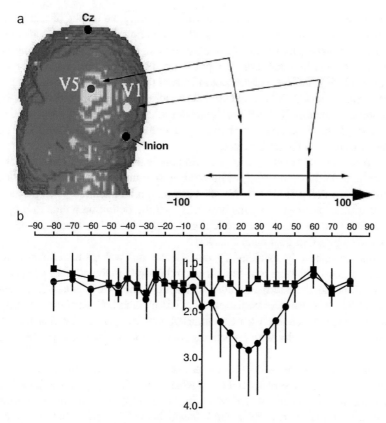

Fig. 3.10 Consciousness manipulation experiments using TMS by Pascual-Leone and Walsh. The dip in the graph starting around +25 ms indicates that TMS stimulation of the primary visual cortex (V1) inhibited hallucinations (motion phosphenes) caused by TMS stimulation of the MT/MST area (V5) (Adapted from Pascual-Leone and Walsh 2001)

was delayed until 25 ms after TMS stimulation of the MT/MST (V5) area. These 25 ms perfectly agree with the time it takes for activity in MT/MST to propagate to the primary visual cortex through top–down neural projections. From this, Pascual-Leone and Walsh concluded that the primary visual cortex is indeed vital to visual consciousness, even in cases where higher-level areas are directly stimulated.

The Difficulty of Using Manipulation Experiments to Search for NCC

Do Pascual-Leone and Walsh's findings then prove that the primary visual cortex falls within NCC? Well, not necessarily.

This case demonstrates the difficulty of interpreting the results of manipulation experiments. Just because we break something and find that doing so leads to some loss of functionality, that does not necessarily imply that the broken thing is the essence of the lost function. For example, we can yank the batteries out of a radio, thereby rendering it silent, but that doesn't mean that batteries constitute the fundamental nature of radios. What makes a radio a radio is a resonance circuit comprising a coil and a condenser. It doesn't really matter what kind of power supply the radio uses—it might have batteries, or a power cord, or even a hand crank—so long as the resonant circuit is in the desired state.

The problem with manipulation experiments is that changing some aspect of a function can produce an effect, even when it was not the fundamental nature of that function which was altered. We must pay particular attention when performing manipulation experiments in pursuit of NCC, whose definition points to irreplaceable neuronal circuitry and dynamics that genuinely generates subjective experience, going one step further than mere causality.

In this sense, strictly speaking, Pascual-Leone and Walsh's experiments cannot exclude the possibility that top–down triggered activity in the primary visual cortex only functions like the battery in a radio, which indeed helps to maintain the necessary activity of NCC somewhere downstream, but it itself does not constitute the "critical" neural circuitry that genuinely generates our subjective experience. Perhaps, the NCC may maintain its activity with an artificial device that replaces the primary visual cortex.

Of course, their observation that the primary visual cortex plays a "causal" role in conscious vision, even in special cases where it is initially bypassed, is extremely interesting and beneficial for understanding the neural mechanisms of consciousness. And of course, there is still a good chance that the primary visual cortex constitutes NCC. Finally, for the record, Pascual-Leone and Walsh themselves, given their fabulous results, do not make the claim that the primary visual cortex is part of NCC.

Process of Elimination Through Manipulation Experiments

Well then, in what cases do manipulation experiments yield decisive conclusions regarding NCC? Returning to the radio metaphor, if we remove a part from the radio—its housing, say, or a manufacturer's label—and it continues to function normally, then we can state with certainty that the part is not included in the radio's fundamental nature. Similarly, if we manipulate specific neurons and that manipulation does not interfere with subjective experience, then we can safely exclude those neurons from NCC. In other words, using manipulation experiments to investigate NCC requires an approach similar to that described in the previous chapter; we must use a process of elimination to whittle away non-NCC factors one at a time.

Such a process of elimination requires extremely precise manipulations that pinpoint the target of interest, because counterintuitively, the goal becomes finding manipulations that have no effect on subjective experience. In the remainder of the chapter, I describe a revolutionary method that allows manipulations at levels of precision that researchers previously could never dream of achieving, and show how this new method can be applied to investigations of NCC.

Optogenetics: A New Development in Manipulation Experiments

The revolutionary method is called optogenetics. Optogenetics refers to embedding artificial ion channels that open or close through light stimulation on neuronal cell membranes of specific neuronal types, allowing us to freely manipulate the activity of target neurons. Interestingly, this definition started out not as a definition, but more as a desire. In 1999, just a few years before his passing, Francis Crick, in a series of lectures at the University of California in San Diego, described the exact idea as his ideal methodology for further pushing the boundaries of neuroscience.

Several years later, Boris Zemelman and Gero Miesenböck performed genetic manipulations that produced light-sensitive ion channels in neurons of fruit flies. Later, Karl Deisseroth greatly simplified the method, making widespread application of optogenetics possible.

I have used optogenetics in my own experiments and was indeed amazed by its ease of use. The procedure is essentially as follows. You open a small hole in a mouse's cranium, insert a glass pipette (a thin glass tube used to inject a drug), and deliver the substance that forms artificial light-sensitive ion channels on cell membranes. Several weeks later, you simply shine light on the brain through an optical fiber.

The primary advantage of optogenetics is its temporal precision and its ability to target and manipulate specific neuronal types. Its temporal precision allows us to trigger or suppress neuronal firings within a millisecond. Regarding neuronal types, there are dozens of them in the cerebral cortex alone, and each plays a distinctive role in neural circuitry. Figure 3.11 shows just a small sampling, where the neuronal type determines the type of neuron it can send output to, as well as the type it can accept input from. Furthermore, it determines the effect of its output, where neurons can be broadly classified as "excitatory" or "inhibitory"; output from excitatory neurons has a positive effect on the receiving neuron, while output from inhibitory neurons has a negative effect.

The temporal precision and targeted manipulation of optogenetics greatly expand the freedom of neural manipulation. The one drawback of optogenetics is that it can only be used to its full potential in a limited variety of animal models. In fact, its ability to target a wide variety of neuronal types is currently limited to a small handful of animals, including fruit flies, zebrafish, mice, and rats. Considering the long and prominent role monkeys have played in visual consciousness research, this is a harsh restriction. And this is the exact reason why I took a leap of faith in transferring my animal model from monkeys to rodents.

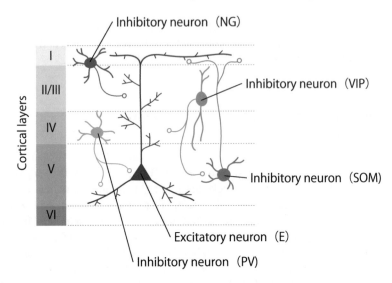

Fig. 3.11 Types of neurons in the cerebral cortex. Excitatory neurons (E: cone cells) and inhibitory neurons (PV: parvalbumin expressing cells, SOM: somatostatin expressing cells, VIP: vasoactive intestinal peptide expressing cells, NG: neurogliaform cells) are shown

Do Rats Experience Binocular Rivalry?

Following the conventional methodology of visual consciousness research, I began the summer of 2011 searching for an optical illusion that could render visual stimuli invisible to rats and mice. My first idea was to invoke binocular rivalry in a manner similar to what Logothetis did with monkeys.

To do so, I spent over six months developing a somewhat complicated device that could be used with rats (Fig. 3.12, top), with the help and support of Johanness Boldt from the machine workshop, Oliver Holder and Theodor Steffen from the electronic workshop, Joachim Werner and Nelson Totah from the Logothetis department, and Damian Wallace from Jason Kerr's group, all at the Max Planck Institute for Biological Cybernetics in Tübingen. Rats were fixed on a freely rotating ball through a metal fitting attached to the skull, and made perceptual reports according to the speed and direction in which they ran. The ball had a passive (rat-directed) mode for measuring the ball direction and speed, and an active (motor-controlled) mode to force a ball direction in an attempt to instruct the correct motor response. To present different stimuli to each eye, I used "parallax barriers" (Fig. 3.12, bottom), where fine vertical stripes placed in front of a monitor led to different rows of pixels entering the left and right eyes.

Although, after all this preparation, and a full year of trial-and-error, my only conclusion was that rats are less than ideal candidates for binocular rivalry. It is the nature of animal experiments that we cannot directly ask subjects what they are experiencing, which makes going beyond the realm of supposition difficult. From

Fig. 3.12 A device developed by the author to invoke binocular rivalry and visual backward masking in rats. A computer mouse is placed atop the device as an indicator of scale. An illustration of a parallax barrier is shown below

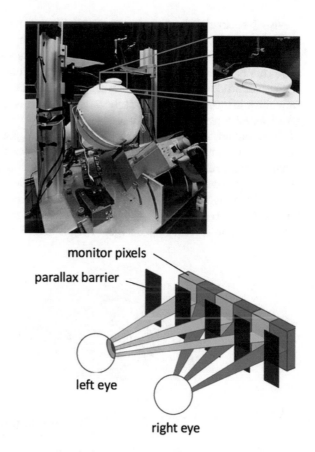

monitor pixels

parallax barrier

left eye

right eye

the behavioral and eye-tracking results, however, I realized the rats were independently moving their right and left eyes (a trick we knew they are capable of), likely allowing them to spatially separate the two supposedly rival stimuli in their visual field, negating the effect of binocular rivalry, say, one stimulus rendering the other invisible.

I had eagerly engaged in this project with aspirations of revisiting Logothetis's work with modern tools allowing more advanced neural manipulation, but one year later I was left with nothing but a deeper understanding of the challenges inherent in establishing a new experimental paradigm.

A New Hope: Visual Backward Masking

There was no point in sulking, however, since I was already committed to this line of research for too long. Having failed with binocular rivalry, I turned to the co-champion of visual illusions that render stimuli invisible, visual backward masking.

Visual backward masking is an optical illusion where an initial stimulus is rendered invisible by a subsequent stimulus. As shown in Fig. 3.13, we first present a low contrast grating (the target stimulus) for a very short time (approximately 16 ms). This is followed by a brief interval, then a second 16-ms presentation of a higher-contrast lattice pattern, which serves as the mask. While there is individual variability, in most human subjects the target stimulus is rendered invisible when the time interval between the two stimuli is less than 100 ms.

Even without neural measurements, the psychological properties of backward masking tell us much about the neural mechanism of visual consciousness. The most straightforward interpretation of how the masking stimulus exerts its effect backward in time is to presume that target neural activity needs to be sustained to enter conscious vision. Taking into consideration the above-mentioned "magic number" of 100 ms for human observers, it needs to be sustained for 100 ms and interfering even only with the last portion renders the target invisible (Fig. 3.13).

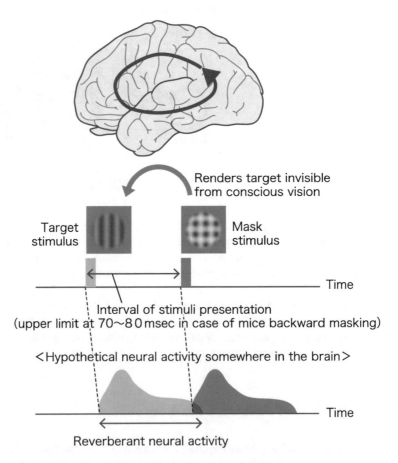

Fig. 3.13 Visual backward masking and reverberant neural activity

Do Rats Experience Visual Backward Masking?

Unlike with binocular rivalry, there was only a small chance that the special way in which rats can independently move their two eyes would interfere with visual backward masking. Even so, there was no guarantee that rats would experience this optical illusion.

The configuration of visual areas in rat brains is quite different from that of human and monkey brains. For this and several other reasons, it was certainly conceivable that visual backward masking would not occur in rats. It was even possible that rats could not perceive the briefly presented low-contrast target stimulus without any masking, as colleagues with much more experience with rat vision pointed out to me. I was thus nearly convinced that I would meet with another failure, even as I forged ahead with the project.

However, the second time proved to be the charm, and I was able to show that rats do indeed experience visual backward masking. I was furthermore able to show that time characteristics were quite similar to those observed in humans; perceptual reports of the target stimulus were significantly reduced when the time interval between presentation of the target and masking stimuli was less than 70 to 80 ms.

At this point, more than two years had passed since I had turned to rats as an animal model for visual consciousness research.

Do Mice Experience Visual Backward Masking?

In the spring of 2014, around one year after obtaining these results, I attended the biennial Tucson Conference for the Science of Consciousness, where I had the opportunity to discuss my findings with Christof Koch. He was astonished and also very pleased when I told him that visual backward masking works in rats, and immediately brought up the possibility of conducting joint research. Koch had recently moved from the California Institute of Technology to the Allen Institute for Brain Science in Seattle, where they have a strict rule to use only mice as their animal model. Guided by the strong views of Paul Allen, the founder of the Allen Institute—and, several decades earlier, the co-founder of Microsoft—the Institute had limited its animal models to mice alone in order to push forward neuroscience in the most effective way. For me, the Allen Institute's state-of-the-art facilities were one attraction, together with the opportunity to discuss and collaborate with another leading figure in the field.

My biggest hesitation was that I would have to cast aside the rat model that I had spent years developing and start over with mice. Success with rats was no guarantee of success with mice. The weight of a mouse brain is only one-tenth that of a rat brain, and their eyesight is less than one-tenth as keen (Fig. 3.14).

Around the same time, I also learned that a group had heard of our success and was performing its own experiments with backward masking. When I presented

Fig. 3.14 Body size of rats versus mice

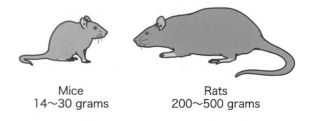

Mice
14~30 grams

Rats
200~500 grams

my rat results at the Society of Neuroscience Conference in 2014, a young student approached me and casually mentioned that his group had just recently begun investigating visual backward masking in rats. I didn't hold this against him—students research whatever their boss tells them to, after all—but I was pretty sure I knew when and how the content of our studies had leaked. I had heard rumors, but this was the moment I realized how damaging leaks could be. I had followed many twists and turns before arriving at my results. Conducting this research was a huge gamble with no assurance of success. In contrast, those following were more or less guaranteed success, and so could devote resources without hesitation.

Investigating NCC Through Optogenetics

In the end, we decided to investigate this new animal model in Tübingen (Fig. 3.15), since it was on my shoulders to prove that backward masking also occurs in mice. I turned to Laura Busse and Steffen Katzner at the University of Tübingen (situated on the other side of the hill from the Max Planck Institute) for help, and they became my new collaborators. Thanks to their experience and expertise working with mice vision, combined with what I had learned working with rats, within six months we

Fig. 3.15 The mouse training setup used by the author, based on a system developed at the Allen Institute by Shawn Olsen and others. The mouse is placed on a freely rotating disc the size of a CD and secured via a metal fitting attached to its skull, and reports its perception through its running speed. The water given as a reward is dispensed from a spout mounted in front of the mouse (left). Several mice can be trained simultaneously (right)

confirmed that mice, too, experience visual backward masking. The "magic number" for mice, namely the longest interval between target and mask for which the illusion retains its effect, was 80 ms, similar to humans, monkeys, and rats.

So finally, everything was set and ready for my manipulation experiments. We could assume that in mice too, neural activity for the target stimulus needs to be sustained for 80 ms to enter conscious vision. The question was, how can we possibly close in on NCC using optogenetics?

We postulated that the magic number could be transferred to the properties of NCC, that NCC should show sustained activity of 80 ms and any disturbance to it would render the target stimulus invisible. By combining the sustained nature of NCC with a special variant of optogenetics, we could explore NCC through a process of elimination.

The special variant is called optogenetic silencing. In this technique, we use optogenetics to forcibly activate a certain type of inhibitory neuron, in our case PV neurons (Fig. 3.11), which are known to have inhibitory connections to all neuronal types throughout the six-layered structure of the neocortex.

As Fig. 3.11 illustrates, the general rule of thumb is that inhibitory neurons in the cerebral cortex have only short-range connections, so all action potentials targeting external visual areas need to be emitted by excitatory neurons. This means that if we can stop those excitatory neurons from firing, we can cut off all outgoing communications to the rest of the brain, turning a targeted area into a kind of neural black hole; action potentials may enter it, but they can never leave. From the perspective of other brain areas, it is as if the optogenetically silenced area had disappeared entirely. Furthermore, the high temporal precision of optogenetics makes it possible to "kill off" these areas instantaneously. As a side note, this is a clear advantage over TMS, where any cortical stimulation, even the ones that exert inhibitory effects as in TMS-induced scotoma, likely leads to scattering of action potentials in neighboring areas.

By combining the behavioral results of visual backward masking with optogenetic silencing, we investigated NCC through a process of elimination, based on the following experimental logic:

(a) From the results of visual backward masking, neural activity constituting NCC should reverberate for 80 ms.

(b) Furthermore, the NCC requirement for "minimal set" and "joint sufficiency" of neural activity demands a negative impact on visual perception of the target stimulus when the NCC reverberant component is suppressed.

(c) However, the converse of (b) is not necessarily true. Specifically, a negative impact on visual perception resulting from optogenetic silencing does not necessarily mean that this particular activity constitutes NCC; it may have only been playing an auxiliary role, as in the previous "battery and radio" argument.

Based on (a) above, if the duration of neural activity evoked by the target stimulus in a particular visual area is less than 80 ms, that area can be excluded from NCC ("1" in Fig. 3.16). From (b), if the reverberant component of neural activity related to

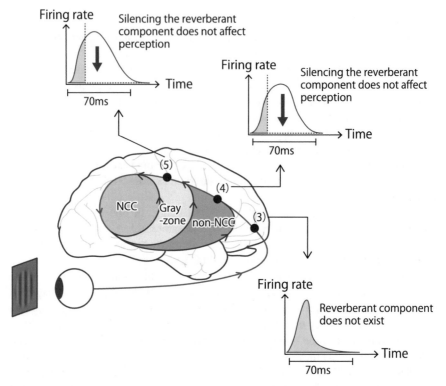

Fig. 3.16 An experiment combining visual backward masking in mice with optogenetic silencing

the target stimulus lasts for more than 80 ms, but there is no effect on visual perception when that reverberant component is silenced through optogenetics, again, that neural activity can be excluded from NCC ("2" in Fig. 3.16). Finally, from (c), if the reverberant component lasts for more than 80 ms and visual perception is negatively affected when that continuous component is silenced through optogenetics, then we cannot determine whether that neural activity is included in NCC, placing it in the gray zone ("3" in Fig. 3.16).

The Future of Neural Manipulation in Search of NCC

My current research follows the experimental logic described above. While we have shown that the primary visual cortex satisfies condition (1) in the figure, where we presented the results in Society of Neuroscience Conference in 2015, we have also preliminary evidence for higher-order visual areas that are categorized as (3) in the figure.

In the future, I intend to first close in on the border between the gray zone and non-NCC regions through a kind of pincer attack from the sides of the primary visual cortex and higher-order areas. However, even if we are successful in roughly determining where that borderline is located in the spatial scale of visual areas, the quest for NCC will still have a long way to go. This is because, as mentioned in Chap. 2, the true borderline will not be at the macro-level of visual areas in the brain, but more likely at the micro-level of individual neurons.

Looking at the bright side, efforts to further improve the spatial precision of optogenetics are underway and researchers have made amazing progress. For example, we can now use lasers and holographic mechanisms to target individual neurons and suppress or activate them separetely. We also have new methods to form artificial ion channels only in recently activated neurons (e.g., neurons that were activated when presented a certain visual stimulus), vastly increasing our freedom in designing manipulation experiments. I consider myself very fortunate to be working in this field and devoting myself to the quest for NCC, in this time of increasingly useful tools for neural manipulation. Nonetheless, as I will explain in the following chapters, I postulate that we need something more than optogenetics, or even the biological brain, for the true understanding of consciousness.

Chapter 4
The Quest for a Natural Law of Consciousness

The Brain as a Neural Network

Before we dive into the true problem of consciousness, I would like to start this chapter with a summary of what I have covered so far. To begin with, we know that individual neurons have limited functionality. They apply weightings to and tally action potentials from other neurons, and when the sum exceeds a certain level, they output their own action potential. These functions are realized through nano-level biological mechanisms such as ion channels and neurotransmitters, and the principles by which these mechanisms operate are more or less understood. It seems unlikely that some unknown mechanism exists that would allow individual neurons to provide the basis for consciousness.

We also largely understand the calculations that neurons perform as a population. There are around 100 billion neurons in a human brain, enmeshed in a huge and complexly intertwined neural network. Calculations take place within this network through the conversion of firing patterns between neuronal groups, determined according to how the wiring between them is routed. In other words, the brain is nothing but an electrical circuit. An inconceivably complex and extensive one, yes, but there is nothing mysterious about the neurons that carry out its elementary processes, nor about the computations that neural groups perform as a network.

Glimpses of Consciousness in Neural Networks

Let's turn next to what experimental approaches to consciousness have uncovered so far. In particular, how does this mysterious electrical circuitry change its operations in accordance with changes in the content of consciousness?

Logothetis's macaque experiments with binocular rivalry have shown that neural activity increases and decreases with the appearance and disappearance of objects in conscious vision. Furthermore, these modulations in neural activity become more

M. Watanabe, *From Biological to Artificial Consciousness*, The Frontiers Collection, https://doi.org/10.1007/978-3-030-91138-6_4

amplified in higher-level visual areas. However, even in the highest visual areas, neural activity never perfectly agrees with the contents of consciousness. While visual experience alternates between perfect existence and nonexistence of visual targets, neural activity never shows such extreme swings. Even when a target stimulus is completely erased from consciousness, neural activity still shows a definite increase compared with conditions where no stimulus is physically presented.

There is thus no part of the cortical system that represents the visual world as-is in our conscious vision. We therefore have no indication of a central system in the brain that is solely responsible for supporting consciousness. Indeed, experimental results suggest that neural activities associated with consciousness and subconsciousness co-exist across multiple hierarchical areas in the visual system. It is unlikely that a clear boundary splits our visual system into distinctly demarked regions of consciousness and subconsciousness at the macro-scale of cortical areas. Rather, a more complex, fractal-like interface appears to be interwoven in multiple visual areas throughout the cortical hierarchy.

We should also keep in mind the insights we obtained from Chap. 3, regarding temporal aspects and neural dynamics of consciousness. Findings suggest that the neural mechanisms supporting consciousness cannot be defined solely as a static set of neurons, but involve more dynamic characteristics such as reverberation and feedback of neural activity.

Is NCC All There Is?

Our understanding of how the brain works as a neural network is deepening every day. Even so, our picture of how the brain supports consciousness remains woefully incomplete. That being said, we have only just begun to study consciousness through scientific experimentation, and the development of new tools such as optogenetics has raised hopes of closing in on NCC.

If this progress does ultimately result in the full identification of NCC, would we have fully elucidated the mechanisms of consciousness? Let's imagine some possible future outcomes of NCC and examine.

Assume we have learned that when neuronal group α in neural network a activates, it produces the sensory experience "red apple," and when neuronal group β activates it produces the sensory experience "green pear" (Fig. 4.1). This neural network a spans visual areas A, B, and C, and neuronal groups α and β produce their respective sensory experiences only when persistently activated for fractions of a second. If we somehow managed to identify both NCCs at such levels, can we conclude that we have perfectly elucidated the mechanisms by which humans become conscious of a red apple or a green pear? Unfortunately, we cannot.

Now imagine that we have also learned the temporal sequence and all causal relations in the associated neuronal firings (Fig. 4.2). In neural network a, neuron subgroup a_1 first activates upon receipt of sensory input, thereby activating neuron subgroup a_2, which in turn activates a_3. Only then does "red apple" rise to the level of

Fig. 4.1 NCC for apples
and pears

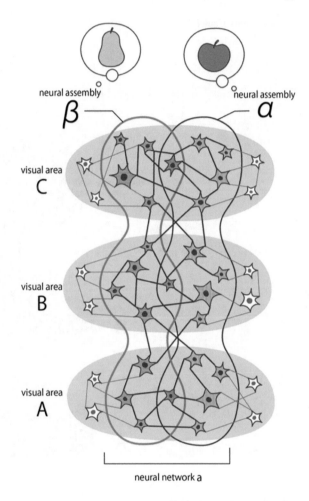

consciousness. If we have learned all of this, can we finally say that we have solved
the mystery of consciousness? The answer, unfortunately, remains "no."

There is no problem if we limit ourselves to asking how the brain identifies an
apple from sensory input and outputs motor commands to reach for and grab it. If
we manage to identify the time series of neuronal firings, causalities therein, and the
calculation principles underlying it all, we would have learned all there is to know
about sensorimotor conversion. Doing so amounts to deciphering how the electronic
circuitry in a robot manages image processing and manipulation of its body.

The nature of consciousness, however, would remain elusive, because we still
would not have learned how subjective experience arises from a chain of neuronal
firings as they convert sensory inputs into motor commands. We would still have no
idea how a pack of neurons in our heads produces the sensations of seeing a bright
red apple, reaching out to grab it, and holding it in our hands.

Fig. 4.2 More detailed NCC
for apples

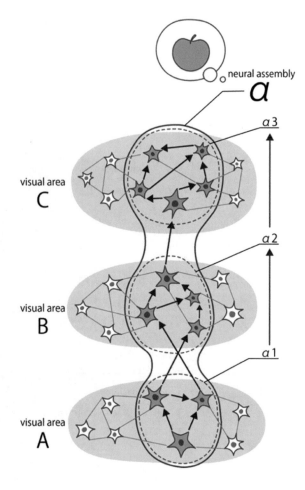

The Mill Argument

The seventeenth-century philosopher Gottfried Leibniz was one of the first to grasp
the fundamental nature of this problem and conducted the following thought experiment. Since the workings of the brain were a complete mystery in his day, he
humorously considered the case of a conscious machine with a size of a mill. It runs
as follows:

"We must confess that perception, and what depends on it, is inexplicable in terms
of mechanical reasons, that is, through shapes and motions. If we imagine that there
is a machine whose structure makes it think, sense, and have perceptions, we could
conceive it enlarged, keeping the same proportions, so that we could enter into it, as
one enters into a mill. Assuming that, when inspecting its interior, we will only find
parts that push one another, and we will never find anything to explain a perception."

Regarding objectivity, we know today that our brains are not all that different from Leibniz's hypothetical machine. Our eyes, ears, and other sensory organs gather information that is passed on to the brain as action potentials. When those signals reach the brain, they propagate through a series of neural networks and are eventually converted into action potentials for controlling individual muscle fibers. The action potentials are transmitted down the spinal cord and drive the motion of our hands and feet. While the brain as an information processing device may be far more complex than Leibniz's machine, in principle, it is fully comprehensible. These objective issues are what the philosopher David Chalmers categorizes as the "easy problem."

The subjective nature of the brain, however, remains unexplained, just as with Leibniz's mill argument. Regardless of how well we elucidate the objective nature of the brain, we will not have taken a single step closer to its subjective nature.

The Hard Problem of Consciousness

This separation between objectivity and subjectivity lies at the heart of the problem of consciousness. The philosopher Joseph Levine calls it the "explanatory gap," and Chalmers calls it the "hard problem."

The objectivity of a neural network is what can be observed by inserting electrodes. In contrast, consciousness can only be subjectively experienced by "us," the protagonist of Descartes' "I think, therefore I am." "We" are the neural network. Subjectivity is nothing more or nothing less than what the neural network is experiencing in first person.

So how does a "pack of neurons" come to possess a first-person perspective of subjective experience? As the American philosopher Thomas Nagel asserted, "an organism has conscious mental states if and only if there is something that it is like to be that organism—something it is like for the organism." So why is there something that it is like to be a neural network? It is hard to imagine that there is anything that it is like to be, for example, a radio, despite the rather complex interaction among countless numbers of electronic parts. Meanwhile, our subjective experiences of seeing, hearing, and thinking unequivocally demonstrate that our functioning brain possesses a first-person perspective, despite being at its basic level, simply electronic interactions between neurons. As mentioned in the first chapter, the mere existence of an "I" in the brain is the biggest mystery.

Objectivity and subjectivity are two sides of the same coin, action potentials passing through neurons and the resulting subjective experience, but we have no clue what binds the two. There must be something in between, some missing link that causally binds the third-party perspective of neuronal activity with the first-person perspective of subjective experience. The mechanism by which one side can flip to the other—the scheme by which the operations of a neural network can be converted into the fine, rich subjective experience when appreciating the vivid red of a rose—for now seems beyond our reach.

The Consciousness of a Thermostat

To highlight the difficulty of the problem, let us examine whether a thermostat might have consciousness and consider what might fill the "gap" between objectivity and subjectivity. While this may sound like a joke, it is actually a topic of hot debate among philosophers.

A thermostat is a simple device that adjusts the output of a heating unit according to the current room temperature (Fig. 4.3). Within it is a band formed from two thin strips of metals with different thermal expansion properties, causing this band to deform as the temperature changes. The band can be used as a switch: when the room is cold, it bends one way, completing a circuit and turning on the heater, and when the room warms up, the band bends the other way, breaking the circuit and turning the heater off.

Here again, the leading figure in the philosophy of consciousness, David Chalmers, makes an appearance, insisting that this thermostat is indeed conscious, at least to some extent. The central tenet of his philosophy is what he calls the "dual-aspect theory of information," namely, that all information has both objective and subjective aspects. He argues that because the thermostat retains information about the room's temperature in the bending of its bimetal band, it possesses some minimal level of subjective sensory experience. Although lacking contracting muscles and pores and such, the thermostat's subjective experience of hot and cold should be no doubt quite different from ours.

You might well ask how a serious debate regarding the consciousness of thermostats could exist today, in the twenty-first century. But it should not be all that surprising—it is simply an indication of how poor our current understanding of consciousness is. With respect to consciousness, the biggest difference between an expert and a non-expert is that the expert knows that we know nearly nothing. This

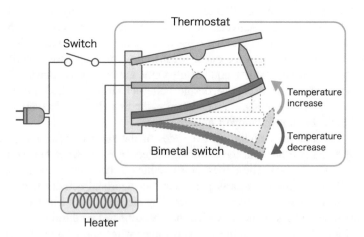

Fig. 4.3 A thermostat

utter lack of knowledge is what currently fills the gap between subjectivity and objectivity. Indeed, because we know nothing, we cannot logically deny any kind of filler for that gap, even Chalmers' claim that "all information contains consciousness."

Of course, if Chalmers is correct, then this book has been wrong from the start, considering that I began by denying the existence of consciousness in the computer I am typing this very book on. According to his theory, there is even consciousness in a rock sitting on the far side of the moon, since like the thermostat, it will expand and contract as the sun periodically heats it up. In doing so, it retains information about its own temperature. In Chalmers's view, there is consciousness literally everywhere.

Can Consciousness Actually Be Understood?

I hope the above discussion has given you a feel for the gap between objectivity and subjectivity. The next question is, can we ever close it?

Some philosophers assert that truly understanding consciousness is impossible in principle. They support this from a variety of rationales, saying for example that a system, our brain in this case, cannot understand itself, or by applying logic so complex as to be opaque to non-philosophers. In contrast, quite a few scientists, especially those outside the field, claim that consciousness can be revealed with neither excess nor deficiency within the framework of conventional science. According to them, issues like the "explanatory gap" and the "hard problem" are not problems at all, and the only feasible statement we can currently make is that we simply have not yet discovered the pathway leading to an understanding of consciousness. Most often, they compare consciousness research with research on life. The mechanisms of life, after all, were completely unknown territory until the early twentieth century; until then, some mysterious force was believed to separate life from non-life. However, countless discoveries in biology over the last century have established that life is nothing more than micro-scale combinations of molecular mechanisms. These researchers argue that just as conventional science was able to pull away the mystical veil that once obscured the nature of life, so too will we eventually unveil consciousness.

Somewhere in the middle are philosophers like Chalmers and John Searle, whom I visited back in 2014 with a half-baked version of my ideas, scientists like Crick and many others—myself included—who spend our days wrestling with the problem of consciousness. While we do not consider consciousness to be seated within some unassailable fortress, we question whether conventional science is sufficiently equipped to scale its walls. The reason for this, as I explain next, is that conventional science is trapped within objectivity.

Since the establishment of modern science, scientists have taken on countless challenges and surmounted many of them. For example, Einstein revealed that mass and energy are equivalent, and that even minuscule amounts of mass can produce huge amounts of energy. The equation $E = mc^2$ that results from his special theory of relativity shows the tight coupling between mass and energy, two aspects of reality that had previously been considered entirely separate phenomena. Watson and Crick's discovery of the double-helix structure of DNA revealed the self-reproducing nature of life to be a function of biomolecules. Their once-in-a-century realization that the exquisite designs that produce life are written in an alphabet consisting of only four letters revolutionized our understanding of life. However, these discoveries only revealed relations between objective phenomena. Mass, energy, self-replication, DNA and even life itself, all of these are aspects of the objective world. Science has thus far only captured phenomena from a third-party perspective, objectivity.

The mission of consciousness science, however, is to bind the objective with the subjective. The subjective comes in the form of subjective experience and poses the question of what a neural network is experiencing in the first person, a new perspective that science has never addressed. In this sense, the science of consciousness is, by definition, a deviation from conventional science.

Box 1: Psychology and Cognitive Neuroscience

In all honesty, it would be misleading to say that current science has never addressed subjective phenomena. Psychology, for example, is situated squarely in the subjective realm. Combining psychology with brain measurements produced the new field of cognitive neuroscience, which has revealed much about relations between subjectivity and neural activity.

How, then, do cognitive neuroscience and the science of consciousness differ in their attempts to address subjectivity? In truth, there is nearly perfect overlap in the people performing this research and the areas of their investigation, at least for the time being. Indeed, when I publish papers, I do so wearing the mask of a cognitive neuroscientist.

However, those strictly studying cognitive neuroscience go to great lengths to avoid mention of the hard problem of consciousness. They use phrases such as "we observe a correlation between the content of consciousness and neural activity" or "when neurons are manipulated, we see causal effects on perception," but they would never go as far as "consciousness arises from observed neural activity."

These scientists are clearly doing their best to ignore the pink elephant in the room. The charitable view is to say that they set aside the hard problem of consciousness because it cannot be addressed within current scientific frameworks. In the journal articles and conference presentations that form the primary battlefields of academia, only philosophers are allowed to address such sticky subjects.

However, the situation is slowly but surely changing, thanks to the quest for the "natural law of consciousness," which I introduce in the following section. Through the search for these laws, we are finally starting to ascend the high walls that have surrounded consciousness for so long.

Natural Laws Are the Foundation of All Science

When I claimed above that the science of consciousness does not fit within conventional scientific frameworks, I did not intend to imply that the two are fully incompatible. Rather, I hold that the science of consciousness needs to adopt the basic logical structure of conventional science.

The key here is natural laws (Fig. 4.4). As I mentioned in the Introduction, natural laws are foundational, meaning that they cannot be derived from scientific theories. Examples include the principle (precisely speaking, the approximation) of universal gravitational force, which describes how two objects attract each other according to their mass and distance, and the principle of the constancy of light velocity, which states that light in a vacuum travels at a fixed speed regardless of the relative motion of the light source and the observer. When asked why these things are true, we can only reply that this is how the universe works.

Natural laws are vital to science. If you trace any scientific theory back to its origins, you will end up at these natural laws. Just as we cannot construct a building without a solid foundation, we could not have built up science without these laws.

If that is the case, we should take it for granted that the science of consciousness requires one too. After all, the science of consciousness is destined to protrude from conventional science and surely needs a new foundation to rest its theories upon. From this standpoint, Chalmers's "Dual Aspect Theory of Information" should be seen as a candidate law of nature, and if it turns out to be true, it will make a superb one of stripped-down simplicity.

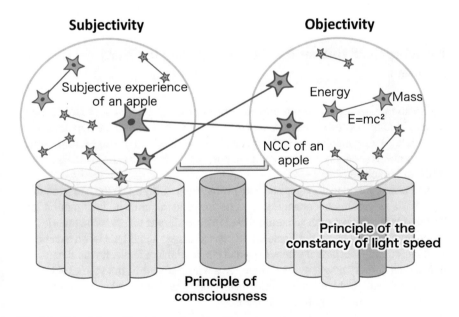

Fig. 4.4 Objectivity, subjectivity, and a natural law of consciousness

Resistance to Natural Laws of Consciousness

Actually, however, Chalmers's proposal is not generally accepted at face value as a natural law, but rather as something of a curiosity. Why? Well, just consider how you felt when I suggested that a thermometer might have consciousness. You probably thought it sounded ridiculous, right? You equate consciousness with your sense of self, so it is difficult to accept that at some fundamental level, you might basically be the same as a thermostat. This is exactly why Chalmers's proposal is not taken seriously, and it is a stumbling block for the science of consciousness. Even scientists, who are supposedly well trained in objective thinking, find it hard to consider themselves objectively; they cannot escape the boundaries of "common sense."

In this respect, I admire Koch, the co-proposer of NCC. In his 2004 book *The Quest for Consciousness: A Neurobiological Approach*, he described Chalmers's hypothesis: "… the audacity of endowing all systems that represent information with experience has a certain appeal and elegance, …." I myself was highly skeptical of the consciousness of thermostats when first presented with the idea, and only half-jokingly brought it up in my university lectures. But when I read Koch's comments, I realized how short-sighted I was. While Koch did not necessarily agree with Chalmers's view, he made it clear that it should be treated as a scientific possibility.

The key to escaping the curse of subjectivity is to realize that the gap separating objectivity from subjectivity works both ways. So long as we have no idea how to connect the two, we must continue to completely exclude conclusions based on "common sense," as science has traditionally done.

Bridging the Explanatory Gap and Solving the Hard Problem

With a nod to Chalmers, accepting the necessity of a natural law would be the first step in transforming consciousness research into a true science, one where hypotheses are formulated upon candidate natural laws and then validated. Interestingly, the moment we embrace its necessity, Levine's "explanatory gap" is bridged and Chalmers's "hard problem" is solved.

I propose the explanatory gap and hard problem exist only under the faulty assumption that a scientific understanding of consciousness requires no new natural law. The reasoning against such an assumption is all too simple: all scientific theories are found upon natural laws, and a science of consciousness should not be an exception. And importantly, because it extends beyond conventional science, it would naturally demand an entirely new law. As soon as we acknowledge this, both the explanatory gap and the hard problem vanish into thin air; the natural law is what bridges the gap and nullifies the difficulty of the problem.

The most crucial aspect of natural laws is that we do not need to theoretically derive them, nor ask why they are the laws. As with the constancy of light velocity, once proven true, there is no point in asking why.

Of course, this does not mean we can propose just anything as a candidate natural law. If the proposal does not meet a certain condition, it cannot become a driving force for advancing the science of consciousness.

That condition is verifiability. If a natural law is not subject to scientific verification, it does not qualify as a foundation for scientific theories. At the risk of offending some philosophers, I believe that this is what separates science from philosophy. It is impossible, for instance, to experimentally verify mind–body dualism, the hypothesis that the material brain is separate from the immaterial mind. Such ideas must therefore remain in the realm of philosophy.

How Do We Verify Natural Laws of Consciousness?

So how can we verify such natural laws? As an example, let's consider how the principle of the constancy of light velocity was verified in 1919 by the British astronomer Sir Arthur Eddington.

At the time, there were two grand theories in physics: Einstein's theory of relativity and Newtonian mechanics. Contrary to Einstein's theory, Newton formulated his theory under the assumption that light velocity behaves like any other, that it would vary depending on the speed and acceleration of the light source and a moving observer. Because they were formulated from different candidate natural laws, the two theories predicted different outcomes for how light is affected by gravitation: Einstein's theory predicted twice as large a bend as did Newton's.

The limited precision of instruments in those days required the use of the most massive object in our solar system, the sun, for any observations of gravitational bending of light. Normally the sun is too bright to allow such measurements, but the darkened sky during a solar eclipse would allow observations of starlight in the vicinity of the sun's rim.

Taking advantage of a total eclipse in 1919, Eddington organized expeditions to Brazil and the island of Principe, off the west coast of Africa. After a delicate process of observation and thorough analysis, Einstein's theory, together with the natural law that it rests upon, held true. When the news hit the world, Einstein's name became known not only among scientists, but to the general public, as we know it today.

The take-home message is, when there are competing candidate natural laws and scientific theories, the only way to validate them is to conduct scientific experimentation and ask mother nature which holds true. Since natural laws form the foundations of theoretical systems, they cannot be verified using the theories they are built upon.

Accordingly, verifying natural laws of consciousness would also require experimental validation, and straightforward reasoning tells us that we should conduct them on brains, the very location in which consciousness resides. But, therein lies the greatest challenge.

Experiments aimed at validating natural laws must eliminate all nonessential elements that can interfere with the results. To demonstrate this, let's consider an apocryphal experiment attributed to Galileo Galilei. Galileo supposedly dropped lead balls of different sizes and weights from the Leaning Tower of Pisa to verify a hypothesis held since the days of Aristotle, that heavier objects fall faster. Everyday experiences actually support this hypothesis. For instance, if we drop a feather and a marble at the same time, the marble will hit the floor first. What Galileo wanted to investigate, however, were the genuine effects of gravity. To do so, he had to exclude all non-gravitational effects. This is why he used two lead balls; air resistance is the non-gravitational effect that he wanted to suppress, and the heaviness of lead served to make its effect negligible in comparison to the effects of gravity. As you may know, both balls hit the ground simultaneously, overturning a well-established theory in Greek philosophy.

The need to exclude interfering effects makes it nearly impossible to apply biological brains to verify natural laws of consciousness. To verify Chalmers's dual aspect theory of information, for example, we would need to extract information—and only information—from the brain. Yet as we saw in Chap. 1, neuronal firing, which is the unit of information, takes place through numerous microbiological mechanisms, and neural information cannot exist without them. We cannot simply exclude them through our experimental setup, as Galileo did with the effects of air resistance.

This issue becomes evident when considering another candidate natural law proposed by Roger Penrose and Stuart Hameroff. They hypothesized that quantum mechanical effects in structures called microtubules are the source of consciousness (Fig. 4.5). Microtubules are tiny structures about 25 nm in diameter that form the

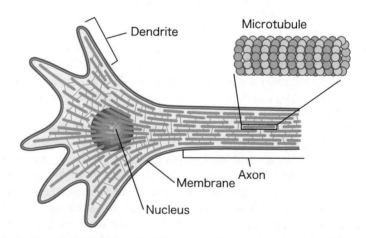

Fig. 4.5 Microtubules in the neuron. These tubular structures have a diameter of about 25 nm (1 nm is one-billionth of a meter). In addition to their role in maintaining the shape of cells and neurons from the inside, they also act as "rails" that carry synaptic vesicles and other entities created in the soma to synapses via axons. While microtubules are small enough that quantum mechanical effects are non-negligible, the question remains open as to whether they are sufficient to support a quantum consciousness theory

cytoskeletons of neurons to maintain their structure and provide platforms for intracellular transport. I did not describe microtubules in Chap. 1 because they do not contribute to neurons' role as information processors.

Many neuroscientists are uncomfortable with this quantum brain theory, because it completely isolates the brain's mechanisms for consciousness from its mechanisms for information processing. However, it is one of the more popular current theories of consciousness, and seems to have ardent fans and detractors in approximately equal numbers. In all honesty, I count myself among the latter, but it nonetheless remains a candidate theory of consciousness. The problem here is that neurons cannot survive without microtubules, so neural information cannot be extracted from potential quantum–mechanical effects of consciousness.

Hence, the biological brain does not qualify as a testbed to ultimately validate the natural law of consciousness.

Analysis by Synthesis

If we cannot validate natural laws of consciousness through experimentation on biological brains, we are left with only one alternative: the use of artificial devices.

Attempts are already underway to provide computers or robots with artificial consciousness. Specifically, this ambitious endeavor seeks to discover the mechanisms by which consciousness arises while attempting to create a new conscious entity. The advantage of this approach is that we can create this entity in whatever way we wish. Just as when humankind learned to fly, we do not necessarily need to follow every example that nature shows us.

The earliest attempts at flight mimicked the flapping wings of birds, but ended in failure. Human breast muscles are far too undeveloped compared with those of birds, so inventors were unable to generate sufficient lift from wings attached to their arms. Next, inventors tried a human-powered helicopter with a spiral rotor (Fig. 4.6). While this design retained the idea of generating downward air pressure as seen in avian wing-flapping, the design itself took a completely different approach. However, it too failed to get off the ground using the power of human muscles. They next looked into large birds like eagles, which fly by gliding rather than flapping their wings. This finally inspired the realization that the upward curve in those birds' wings was key to realizing human flight. This led to fixed-wing gliders, and finally to the powered flight that the Wright brothers achieved. These many attempts at flight not only opened the skies to human travel, they also led to new scientific theories such as fluid mechanics.

This example contrasts two approaches to scientific experimentation: subtraction from nature as we saw in the previous section (e.g., negating air resistance by use of lead balls) and analysis by synthesis, in which we start adding from zero (e.g., the invention of flying machines).

Fig. 4.6 A flying machine design by Leonardo da Vinci

Fading Qualia

Before we discuss building a new conscious entity from scratch, we must address a fundamental question: Is machine consciousness even possible? Today, many scientists and philosophers agree that, in principle, it is. One reason for this consensus is our ever-increasing understanding of individual neurons and how they behave. Roughly speaking, we have laid their functions bare. Of course, our brain includes tens of billions of intertwined neurons, and we are still a long way from fully understanding how they all interact. But even without complete knowledge, we can still consider the possibility of machine consciousness through the following thought experiment, yet another contribution from Chalmers.

First, imagine looking at an apple placed in front of you. As you are subjectively experiencing this apple, we replace a neuron in your brain with an artificial neuron. As we have shown in the previous chapters, there is nothing mystical about how neurons operate, so it certainly seems plausible that someday we will have a silicon device that completely mimics the input–output characteristics of a biological neuron. Assuming the original neural wiring is maintained, other neurons connected to the silicon neuron, and thus the rest of the brain, should not be affected by the replacement.

What would happen if we replaced another neuron, then another and another, until every biological neuron was replaced with an artificial one? Would your subjective experience of the apple disappear, despite the neural network in your head working exactly as it had before? If so, would it suddenly vanish with the replacement of a single neuron that pushed the biological-to-artificial ratio beyond some critical value? Or would the apple gradually fade away little by little with each replacement, until it completely disappeared?

Chalmers ironically named this thought experiment the "fading qualia" and postulates that both cases are unlikely. He concludes that the subjective experience arising from looking at the apple would not fade even after you had a wholly artificial brain. If so, consciousness can reside in machines.

An interesting corollary to this line of reasoning is that our artificial neurons do not necessarily need to reproduce all aspects of biological ones. So long as other neurons cannot differentiate between which kind they are attached to, we can take all the shortcuts we like. For example, our artificial neurons do not need to involve ion channels in their production of action potentials. So long as they faithfully reproduce the input–output characteristics of a biological neuron, their presence should never be detected.

This thought experiment presents many interesting implications. If the assumptions and inferences we made above are correct, then consciousness can exist in an artificial neural network comprising highly abstracted neurons. Suddenly, machine consciousness does not feel so much like a far-off dream.

Digital Fading Qualia

Following Chalmers's reasoning, consciousness can exist in silicon hardware that sufficiently mimics our brain. Although needless to say, this would be an extremely advanced piece of hardware. Sure enough, the neurons can be highly abstracted, but in regard to connectivity between them, it would fully replicate the original brain.

There are around a hundred billion neurons in a human brain, each receiving input from thousands of other neurons and outputting to thousands more. Barring some dramatic development in semiconductor technology, it will be impossible to replicate the scale and complexity of the human brain in hardware. As things stand, we are still struggling at the very fundamental level of realizing three-dimensional wiring in silicon hardware. Well then, can consciousness reside in an artificial neural network simulated on a digital computer?

Unfortunately, we cannot apply the results of the fading qualia thought experiment as-is to these simulations. When simulating a biological neural network on a von Neumann machine (a type of computer in which a small number of central processing units perform all calculations), we must calculate the behavior of neurons one by one in a time-sharing manner. At any given instance, only a limited number of virtual neurons are actually being simulated, unlike in an actual brain or hardware equivalent, where all neurons physically exist and function in hyper-parallel.

So let's next examine whether consciousness may reside in a neural network simulated on a digital computer, by revising Chalmers's fading qualia (Fig. 4.7). We will assume that its computing power is sufficient for simulating the full human brain in real-time.

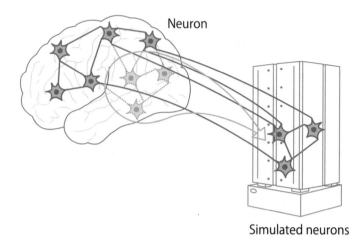

Fig. 4.7   Digital fading qualia

As in Chalmers's original thought experiment, we begin with a fully biological brain. Our first step is to replace one of its neurons with a computer simulation. To do so, we assume a brain-machine interface capable of fully retaining the original synaptic connectivity. So long as these thousands of connections are faithfully reproduced and the neuron's input–output properties are fully replicated, no other neurons remaining in the brain should be affected by this first replacement, just as in the original fading qualia thought experiment.

However, differences can arise with the second replacement, in particular, if the second neuron targeted for replacement was originally connected to the first. Even so, the basic requirements for the fading qualia problem can be fulfilled; if both replaced neurons are perfectly reproduced in the computer, including the interactions between them, the remainder of the brain will not be affected.

In this way, we continue moving neurons into the computer one by one, still satisfying the constraints of the original fading qualia experiment. As in the in-vivo replacement of biological neurons with silicon neurons, it is hard to imagine either a sudden shutoff or gradual fading of qualia during this process. So even after all neurons have been incorporated into the computer, the consciousness that resided in the biological brain should continue on, existing in a simulated spiking neural network contained wholly within the computer.

Box 2: Predicting Natural Laws from Both Types of Fading Qualia

If the fading qualia argument is correct, microbiological mechanisms within neurons do not at all contribute to the formation of consciousness. At the level of individual neurons, the fundamental nature of consciousness relies only on their input–output characteristics. Accordingly, any natural law of consciousness will likely describe highly abstracted causality, such as the firing of myriad neurons triggering the firing of myriad other neurons. Moreover, if the digital version of fading qualia holds, then that natural law of consciousness will address even more abstract phenomena.

There are no physical interactions in a neural network running on a computer. To get a firm picture of this, let's examine the process in more detail.

When a given neuron fires, the computer calculates which neurons the signal will arrive at, and when. After that, time proceeds within the computer, and when it determines that the signal has arrived at the target neuron, it begins calculating the resulting synaptic response. In other words, the firing of one neuron and the resulting synaptic response in the other occur at disjointed times, and there is no chain of physical interactions filling up the time between the two events. Computer-simulated neural networks differ from actual brains in that they only model causality in a highly abstract manner. If we presume the digital fading qualia thought experiment holds, then the natural law of consciousness should be applicable to a form of causality that is symbolic and highly abstract.

Let's dig a bit deeper into this. In the final stage of the digital fading qualia experiment, all neurons have been translated into the computer, and a state of consciousness has been established therein. What happens if we increase or decrease the computer's processing speed? The discontinuous causality described above does not rely on computational speeds, so the existence of the contained consciousness will be completely unaffected. What will change, however, is the relative time with the outer world that this consciousness experiences, as illustrated in the 1994 novel *Permutation City* by Greg Egan. This gets very interesting when, for example, we make the computer's calculations very slow. Will any natural laws of consciousness still hold when time has been slowed to the point of nearly stopping? If so, what does that say about the characteristics of such laws?

Here's another point to consider regarding what natural laws of consciousness should look like. Assume we use some method to record the timing of all neuronal firings within a consciousness-containing neural network. Those readings might come from a brain, or they might come from a conscious artificial network developed at some point in the future. We then use this recorded data to specify the timing at which a collection of simulated neuron-like objects "fire." This neuron suite is equal in scale to the original brain or digital network, but has no connecting elements (synapses, axons) or means to communicate (generation of action potentials) as a network. We are therefore reproducing all neural firings, but only as a flickering of triggered neurons with no causal links among those events.

> Would consciousness exist in such a device? I cannot believe that it does. We can thus infer that a natural law of consciousness is likely to involve chains of causality.

Philosophical Zombies Hinder Tests of Machine Consciousness

One major problem hinders our coming to understand consciousness through analysis by synthesis: we have no methods by which to test for consciousness in machines. This makes our efforts something like developing an airplane on our airless moon: even if we come up with something, we don't have the means to test whether it really works. Analysis by synthesis critically relies on such validation.

The difficulty in testing machine consciousness is that we must account for philosophical zombies, yet another concept proposed by Chalmers. Don't confuse these with the shambling undead you see in movies and television shows. A philosophical zombie is indistinguishable from a normal human being in terms of its appearance or behavior; the only difference is that a philosophical zombie lacks consciousness.

In fact, robots are probably a better analog than zombies in this modern era, and the same concept could be applied to some future robot that is extremely human-like, but without subjective experience. You could even ask such a robot if it has consciousness and it would confidently reply that it does, but, to borrow Chalmers's description, "all is dark inside." Philosophical zombies are an imaginary construct, but a necessary one for working out the constraints of testing machine consciousness.

By this point, some of my readers might have guessed that some variant of the Turing test should be sufficient. For those not familiar with it, this was a test proposed by Alan Turing, the father of modern computing, as a way of determining whether we have achieved artificial intelligence (AI). To avoid identification through appearance or voice characteristics, an investigator communicates indirectly with a computer through the use of a screen and keyboard, and the computer is deemed to have realized AI if the investigator cannot determine whether what's on the other side of the terminal is human or machine.

The Ridley Scott film *Blade Runner*, based on the Philip K. Dick novel *Do Androids Dream of Electric Sheep?*, opens with an investigator conducting a variant of the Turing test on a "replicant." Replicants are androids fabricated largely from biomaterials, making them indistinguishable from humans by appearance alone. In the movie, the test takes the form of an interview, during which the investigator monitors factors such as changes in the subject's pupil dilation and perspiration. Further on in the movie, a replicant comes very close to passing these tests.

The movie never reveals whether replicants have true consciousness. However, it is not difficult to imagine that a machine built so elaborately might pass all these tests, even if they are "all dark inside." So we must conclude that while variants of the Turing test might validate artificial intelligence, they cannot be used for testing consciousness.

Having arrived at the concept of philosophical zombies, we now find ourselves facing a terrifying fact: we must doubt the consciousness not only of futuristic androids, but even of our human neighbors. We can be assured only of that consciousness which arises through our own sensory experiences. Forget machines—so far as I'm concerned, Descartes's "I think, therefore I am" does not even apply to you!

Together, as Leibnitz's mill argument demonstrated, no matter how fully we investigate the inner workings of a device, we cannot determine whether consciousness resides within it. It is thus impossible to determine whether a machine has consciousness from a third-party perspective—in other words, objectively.

Seeing for Ourselves

If it is impossible to objectively test for machine consciousness through external observation or investigation of internal mechanisms, only one method remains: making use of subjectivity. We need to connect our own brains to the machine and "see" for ourselves whether consciousness resides within.

One critical issue is how we connect the machine to our own brains. Not just any connection will do. For example, we already have devices such as artificial retina (Fig. 4.8) and cochlea that produce subjective sensory experiences, but surely, these are not indicators of consciousness in such devices; they are simply substitutes for biological sensory organs, providing alternative sensory inputs to primary sensory areas.

Generally speaking, a subjective experience generated by a device connected to our brain does not necessarily indicate the presence of consciousness within that device. What we therefore need is some way to connect the device so that we obtain subjective experience only when consciousness actually resides in it. We turn to neurophysiology and neuroanatomy to seek for such connections.

Two Hemispheres, Two Consciousnesses

We start by looking at Roger Sperry's research on split-brains, for which he received the 1981 Nobel Prize in Physiology or Medicine. A "split-brain" is a brain that has undergone a procedure called a corpus callosotomy, in which a bundle of nerves connecting the right and left hemispheres is severed.

Like most animals, humans have bilaterally symmetrical brains. As a rule, structures throughout the brain come in left–right pairs. As Fig. 4.9 shows, each pair is split between left and right hemispherical structures, so each set of parts is collectively called a brain hemisphere. The pineal gland that Descartes' dualism held to be

Fig. 4.8 An artificial retina directly connected to the primary visual cortex (from Reuters/Aflo)

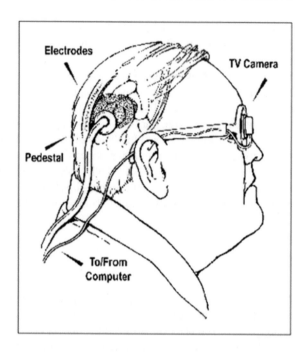

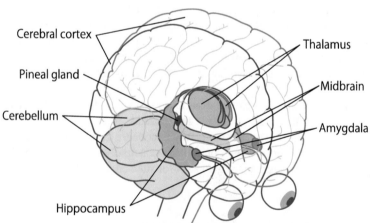

Fig. 4.9 Most brain parts come in pairs, one on each side. The pineal gland is one of the few exceptions

the point of interaction between the physical body and the immaterial mind is one of the few exceptions, in that it sits between the two brain hemispheres.

The right and left hemispheres respectively sense and control the left and right sides of the body (Fig. 4.10). Regarding vision, drawing a vertical line that evenly splits the visual field (the "vertical meridian" in the figure), the right side is processed

Fig. 4.10 Contralateral dominance in the brain. The right hemisphere is responsible for various aspects of the left side of the body, and vice versa, including vision, tactile sense, and motor control. This occurs due to the left–right crossings of nerves in the medulla oblongata, spinal cord, and other locations

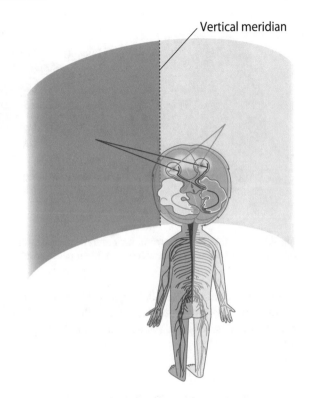

Vertical meridian

by the left hemisphere and the left side by the right. Likewise, skin sensations and motor control in the right and left sides of the body are the responsibility of opposing brain hemispheres.

The right and left cerebral cortices are connected by three nerve fiber bundles: the anterior commissure, the posterior commissure, and the corpus callosum (Fig. 4.11). In standard corpus callosotomy, all three are surgically severed. The procedure is normally performed to treat severe epilepsy, a disease that causes an abnormal increase in neuronal activity.The procedure alleviates symptoms by restricting unusual activity that would otherwise span both hemispheres. While an effective treatment, it inevitably entails after-effects, sometimes so severe that daily life becomes difficult. Examples include the patient's right hand fastening a button, immediately followed by the left hand unbuttoning it, or a knife-wielding left hand preventing a fork-wielding right hand from carrying food to the mouth.

Curiously, when patients are asked about these strange behaviors, they attribute them to something like an alien entity. They claim that their left hand was working on its own in unbuttoning their shirt, or in preventing them from eating their steak. They always seem to relay the left hemisphere's point of view, never the right's. This apparently happens because speech and language centers in the brain are concentrated in the left hemisphere, enabling it to communicate with us.

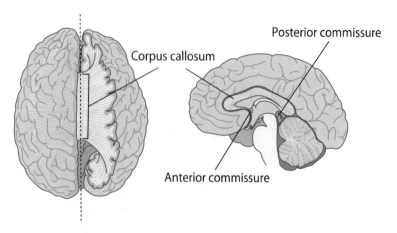

Fig. 4.11 The right and left cerebral cortices are connected by three nerve fiber bundles: the anterior commissure, the posterior commissure, and the corpus callosum

From these patients' left-hemisphere descriptions, it almost sounds like a complete stranger has taken over the left side of their body. It is as if separate, independent streams of consciousness reside in the right hemisphere, from the viewpoint of the left.

Sperry Extracts the Right-Hemisphere's Testimony

It was the experimental validation of two consciousnesses in split-brain patients that won Sperry the Nobel Prize. The challenge in doing so was that, while the left hemisphere was able to use its language centers to eloquently communicate with the outside world, the right hemisphere remained trapped in silence. If Sperry had only obtained testimony from one of these consciousnesses, academia would not have recognized the existence of the other, citing lack of evidence.

In an effort to desperately communicate with the right hemisphere, Sperry focused on the left hand, which was under its control. Perhaps, he thought, we could communicate with the right hemisphere by having the left hand grasp objects. Fortunately for Sperry, the right brain commands sufficient language ability to understand experimental instructions.

Figure 4.12 shows the setup for Sperry's experiments. Split-brain patients sat at a desk with their gaze fixed on the center of a screen in front of them. This allowed the patient's right hemisphere to see only what was shown on the left side of the screen, and the left hemisphere to see only the right. Sperry displayed images of objects on the screen and placed the actual objects on the far side of the desk. He then instructed subjects to either orally describe what they saw on the screen or to use their left hand to grab the presented object.

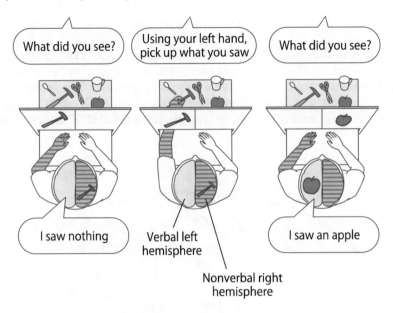

Fig. 4.12 Sperry's experiments on split-brain patients

The results were as shown in the figure. As expected, subjects verbally reported what was displayed on the right side of the screen and grabbed what was displayed on the left. Much more critical was what happened when Sperry displayed objects on the left side of the screen and asked the subject to verbally describe what they saw (Fig. 4.12, left). Strongly supporting the hypothesis that two separate, noncommunicating streams of consciousness reside in the cortical hemispheres, subjects verbally reported that they could not see anything.

Interhemispheric Communication in Cortical Hierarchy

While Sperry's experiments were conducted on split-brain patients, they imply that consciousness in general is not as monolithic a phenomenon as it appears to be. It raises the question of how two potentially separate consciousnesses are bound together in non-split-brains.

Inspired by the results of Sperry's split-brain experiments, many scientists have analyzed in detail the information passing between the right and left cortical hemispheres. The most rigorous method for doing so is to surgically remove a monkey's corpus callosum, euthanize it after a few days, and then examine its brain. The delay of euthanasia is to allow decay of synapses whose axon was severed and detached from the cell body. Through observations of these necrotic synapses, we can trace exactly where synaptic projections from the opposite hemisphere land on the cortical surface.

Figure 4.13 shows an image that was painstakingly created by observing thin slices of an extracted brain under a microscope to visually pinpoint the positions of necrotic synapses. The first point to note is the stark contrast in the concentration of necrotic synapses; some regions have no connections at all, while others are densely packed. Focusing on the densely packed regions, the box labeled "a" in the top image

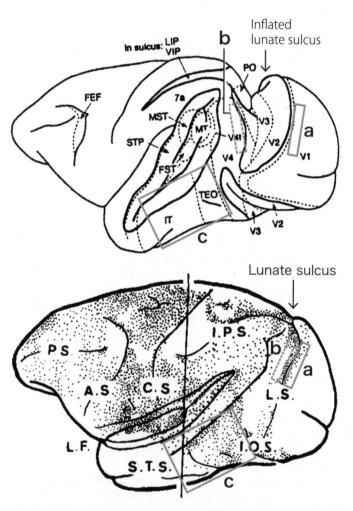

Fig. 4.13 Neural connections between hemispheres. (Top) A monkey brain with several grooves (sulci) spread to depict the hierarchical structure of visual areas. (Bottom) A similar image showing necrotic synapses (black dots) following cutting of the corpus callosum (top: Adapted from Gross et al. 1993, Copyright ©1993 Society for Industrial and Applied Mathematics. Reprinted with permission. All rights reserved. Bottom: Reprinted from Pandya et al. 1971, Copyright (1971), with permission from Elsevier)

in Fig. 4.13 shows the boundary between the primary and secondary visual areas, and the box labeled "b" is the boundary between the third and fourth visual areas. In terms of retinotopy, these regions coincide with the boundary between the left and right visual fields, the vertical meridian. In contrast, there are very few necrotic synapses in the surrounding regions. This suggests that in the low- to mid-level visual areas comprised by the primary through fourth visual cortices, interhemispheric connectivity exists only to minimally stitch together the left and right visual fields (Fig. 4.14). This characteristic spatial bias in interhemispheric connectivity ceases to exist in higher visual areas. The area labeled "c" in Fig. 4.13 corresponds to the inferior temporal (IT) cortex, the highest-level area in the ventral pathway, where we find specialized neurons such as those for faces. Here, necrotic synapses are much more broadly distributed, demonstrating that higher areas possess interhemispheric connectivity that covers the entire field of view (Fig. 4.14).

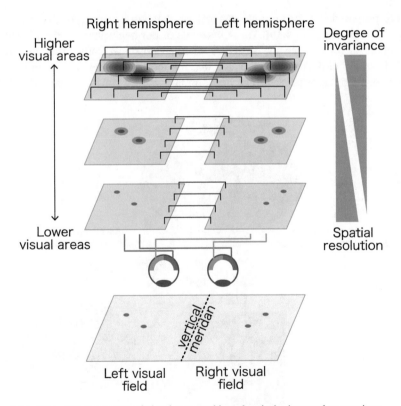

Fig. 4.14 Hierarchical structure of visual areas and inter-hemispheric neural connections

A Subjective Test for Machine Consciousness

Both the experiments on split-brain patients by Sperry and the physiological findings of interhemispheric connectivity point to one fact: regarding visual consciousness, the left and right hemispheres are equals. Namely, there is no asymmetric relationship in which one hemisphere generates consciousness, while the other just provides it with visual information. There is simply not enough interhemispheric neural connectivity in low- to mid-level visual areas to communicate high-resolution visual information between the two. Even if there were, neither hemisphere has sufficient capacity to process the extra input. This is portrayed in the fact that the visual information represented in the left and right hemispheres are strictly exclusive; the right hemisphere holds only information of the left visual field, and vice versa. We can thus conclude that the two hemispheres form a primary–primary configuration in terms of conscious vision.

My proposed test for machine consciousness takes advantage of this primary–primary construct (Fig. 4.15). We replace one of our hemispheres with a mechanical hemisphere, and see for ourselves whether we subjectively experience integrated visual fields, including the side that the mechanical hemisphere processes. If we do, from the primary–primary constraint, we must conclude that a stream of

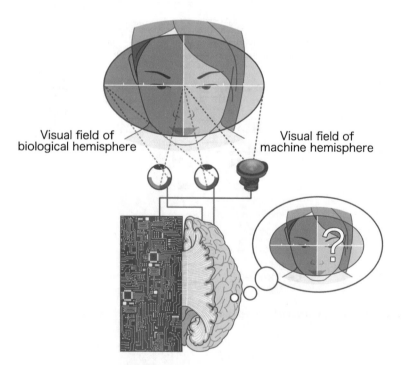

Fig. 4.15 A subjective test for machine consciousness

visual consciousness has emerged in the mechanical hemisphere, and that it is linked to our own biological stream of consciousness.

The key to this test will be maintaining the biological constraints on interhemispheric connectivity when connecting the mechanical and biological halves. Namely, at mid- and low-level visual areas, the only interhemispheric neural connections should be those that bridge the boundary between the left and right visual fields. If we maintain this restriction on connectivity between the mechanical and biological hemispheres, it will be impossible for the biological half to treat the mechanical half as a secondary system that simply provides high-resolution visual input, and furthermore, act alone in generating the subjective experience of both visual hemifields. If there are any doubts, we may record neural activity and validate whether the biological hemisphere underwent any metamorphosis (e.g. low-mid-level areas expanding its visual field coverage to the machine side of view). Confirming that no such transformation has taken place would leave no room to doubt that consciousness resides in the machine.

Having read so far, some of you might suspect that, because we would not have the technological means to administer the proposed test anytime in the near future, all of this, including the subsequent mind uploading, is just fantasy or science fiction, or at best, yet another thought experiment. But wait until the final chapter before drawing your hasty conclusion. I wouldn't be spending my time writing a whole monograph if I had no hope for materializing it while I still have the chance!

Chapter 5
Is Consciousness Information or Algorithm?

Candidate Natural Laws of Consciousness

As described in the previous chapter, the natural law of consciousness is what connects the objective with the subjective. While we may experimentally validate whether proposed natural laws of consciousness are truly laws, there is no point in questioning why they are the laws; as with the principle of the constancy of light velocity and other established natural laws, all we can do is shrug and say that's how our universe works. In a sense, scientists and philosophers are free to indulge in proposing these natural laws without providing rigorous theoretical backup, which by definition is simply not possible.

Candidate natural laws of consciousness will unquestionably tie together the objective and subjective aspects of consciousness (Fig. 4.4). The subjective side of consciousness is straightforward and there is not much to choose from, only our own subjective experiences. So the challenge in proposing laws for consciousness lies in selecting the appropriate objective side. Let's look at two examples.

Chalmers' dual aspect theory of information and Tononi's integrated information theory, which I explain next, take information as the objective side. The difference between the two lies in how they delineate the type of applicable information: while Chalmers considers consciousness to exist in all information, Tononi asserts that it exists only in information in a particular state, which he describes as "integrated."

The first step to understanding Tononi's integrated information theory is grasping what he means by non-integrated information. This is perhaps best done through the example of a camera sensor.

The sensor in a digital camera has rows and columns of three types of pixels: red, green, and blue (Fig. 5.1). Each pixel element accepts and measures light from a filter that only allows light of a specific wavelength to pass through. In this way, the sensor records the visual world that passes through the camera lens. Combinations of high-performance lenses can perform this task at levels far surpassing anything that a human eye is capable of. Even so, is the camera really "seeing" the visual world?

© Springer Nature Switzerland AG 2022
M. Watanabe, *From Biological to Artificial Consciousness*, The Frontiers Collection, https://doi.org/10.1007/978-3-030-91138-6_5

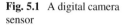

Fig. 5.1 A digital camera sensor

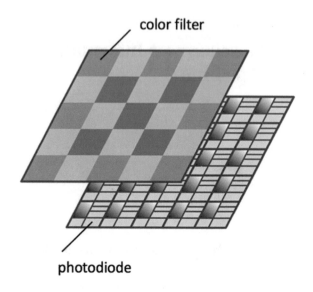

Integrated information theory holds that these sensors do not truly see the world. This theory has many ardent supporters, in part because—unlike Chalmers' theory—it allows us to mathematically derive conclusions that agree with our intuition (though I hope you will forgive me for skipping the actual math in this book).

According to integrated information theory, consciousness does not reside in these sensors because the information they contain exists only in fragments. The amount of light that an individual pixel receives depends exclusively on the object in front of the lens, and that pixel exerts no internal interaction with any other pixel. In contrast, our conscious visual experiences are integrated into a single image so that all scenery visible to our eyes is perceived as a cohesive whole. Through this straightforward reasoning, applying constraints derived from the nature of our subjective experience, integrated information theory denies that sensors with only fragmented information possess consciousness as we know it.

Integrated information theory is compelling in its denial that conscious sensory experience can reside in disjointed information. The part more difficult to comprehend is what the theory defines as integrated information. In the next section, I will focus on its very original definition.

Integrated Information Theory

If the disjointed information in camera sensors is not integrated, then what is? Tononi and his colleagues define integrated information as that in which the "whole is greater than the sum of its parts." In the sensor example, individual fragments of pixel information are independent of one another, so the amount of information the sensor holds is exactly the sum of the information held by its individual pixels, and thus that information does not fit this definition.

The mathematician David Balduzzi, a good friend of mine from our years together at the Max Planck Institute Tübingen, made significant contributions to the early establishment of this theory. Below, I present a description of integrated information as he first explained it to me in our weekly discussion session we held at the institute cafe, also as described in their first publication.

Figure 5.2 shows two observation values and two neurons. The observations are of the size and brightness of squares. Each of the two neurons emits action potentials when its observation values fall within a specific range. Regions a and b in the graph at the upper right show the respective range of the preferences of neurons A and B. Specifically, neuron A fires when squares are small and dark, while neuron B fires when squares are large and bright. Furthermore, the areas of regions a and b are each exactly half the area of the overall range.

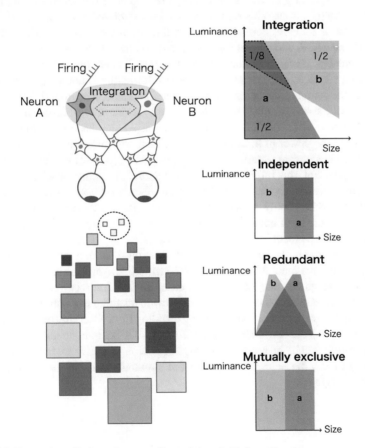

Fig. 5.2 Integration of information according to integrated information theory

Let's consider this situation using a mathematical concept called mutual information, which describes an increase in knowledge about a variable of interest resulting from the observation of another variable. In our example, neuronal firings are the observed variable, and the range of properties of the square is the variable of interest. As a specific example (Fig. 5.2 top right panel), let's consider the mutual information resulting from the observation of neuron A firing. Before this neuron fires, we know nothing of the square's brightness or shape, both of which could be anything we see in the whole region. When neuron A fires, however, the range of possible brightness and size is reduced to half. Using bits as our unit of measure, we can say that this narrowing of possibilities to half the previous range represents a one-bit increase in mutual information. We also gain one bit of mutual information when we observe neuron B firing, since it too halves the range of possibilities.

Next, consider the case where neurons A and B fire simultaneously. This would only occur when the square satisfies both criteria, something that happens only in a narrow range of brightness and size: only one-eighth of the squares in the image will trigger both neurons. This results in three bits of mutual information ($1/8 = 1/2$ to the power of 3). Now, comparing the whole with the sum of its parts—namely, one bit for neuron A and one bit for neuron B, two bits in total—we can see that the whole is greater. The information is thus "integrated" as per our definition.

Now, let's look at three other cases where information is not integrated. As we have seen, the combination of response characteristics among neurons determines the outcome. In the second graph from the top on the right side of Fig. 5.2, the information obtained from the observation of both neurons firing simultaneously, the "whole" ($1/4 = 1/2$ to the power of 2; 2bits) is equivalent to the sum of mutual information obtained by independent observation, "sum of its parts" (1bit + 1bit = 2bits). Because they are equivalent, this information is not integrated. Looking closely, we can see that neuron A has no interest in the brightness of squares; it responds only to size. Conversely, neuron B only responds to brightness. As in the case of camera sensor pixels, the responses of neurons are not affected by each other, so the information they carry is independent in this particular case.

In the third graph, the sum of mutual information separately obtained from two neurons exceeds the mutual information of the whole, so again, the information from the two neurons is not integrated. In this case, information received from the two neurons is redundant; in other words, the narrowing resulting from observing neuron A firing largely overlaps with the narrowing resulting from observing neuron B firing. Generally, in such cases of redundancy, information is not integrated.

In the fourth graph, the range-narrowings of neurons A and B are mutually exclusive. In this case, the neurons will never fire simultaneously, so their information cannot be integrated.

Preliminary Validation with the Proposed Test for Machine Consciousness

Now that we have a basic understanding of the two candidate natural laws that take information as the objective side, let's attempt to verify them using the line of thinking of the subjective test for machine consciousness, which I proposed in the previous chapter. As discussed, however, natural laws can be only validated through actual experimentation, so the following attempt is solely for the purpose of obtaining insights for further discussion.

We begin with Chalmers' double-aspect theory of information. As Koch stated, there is an appeal to the simplicity of Chalmers' proposal that consciousness resides in every information, from human minds to moon rocks, but unfortunately, because of this exact reason, the subjective test for machine consciousness would likely suffer. While Chalmers asserts that information produces consciousness, he does not propose how differences in qualities or quantities of information can result in anything from the primitive consciousness of a stone to our own presumably more advanced consciousness. This is problematic because any candidate artificial brain hemisphere we might build will have some minimal amount of information. Thus, it is not obvious how we may design an experiment following Chalmers' hypothesis in which one type of artificial hemisphere should yield consciousness and another should not; the theory is so all-encompassing that there are no cracks in which to wedge the constraints imposed by a two-hemisphere brain. Here I cite the full version of Koch's statement on Chalmers' view, "While the audacity of endowing all systems that represent information with experience has a certain appeal and elegance, it is not clear to me how Chalmers' hypothesis can be tested scientifically," which I must agree with.

Regarding Tononi's integrated information theory, we will see whether information within our two biological hemispheres is "integrated," as defined in the original version of integrated information theory.

First, we'll consider mid- and low-level visual areas in the brain. Neurons in these visual areas have small receptive fields, with those in the right hemisphere responsible for the left visual field and vice versa. There is almost no overlap in their respective ranges. Visual information in the hemispheres is thus exclusive, so it is likely that the information generated by them is not integrated at this level.

What about high-level visual areas? In these areas, the receptive fields of individual neurons are much larger than in low- and mid-level neurons, and many cross the vertical meridian and extend to the opposite side. In terms of receptive fields, therefore, information contained in high-level areas is redundant. Moreover, there are no reports that the left and right hemispheres have idiosyncratic ways of representing visual information in these areas. Therefore, high-level visual information in the two brain hemispheres can be considered highly redundant, another strike against the potential for its integration as defined by integrated information theory.

At face value, then, integrated information theory in its original form seems unlikely to explain the unification of consciousness between right and left cortical hemispheres. It does not seem to capture the mechanism by which two biological hemispheres combine to form our subjective experience of a unified full field vision. Note, however, that up to this point I have only described what is called "version 1.0" of integrated information theory. It is being extensively developed by countless researchers, and we need to keep an eye on how it addresses the problem.

I must confess, however, that I see more a general problem in assigning information as the objective side for natural laws of consciousness.

The Problem with Equating Information and Consciousness

The biggest problem with equating information and consciousness is that information alone has no meaning. Meaning comes only through interpretation.

Let's take a step back and consider the meaning of meaning. The information that a computer manipulates is, in the end, nothing but strings of ones and zeros. Only when software interprets those binary strings, do they take on meaning, in the form of sounds, or images, etc. (Fig. 5.3). To interpret information as sound, software breaks a one-dimensional array of ones and zeros into a number of chunks depending on the format (e.g., signed 16-bit integers or 64-bit double-precision floating-point values), converts them into scalar values, and lines those values up in two one-dimensional arrays for stereo playback. When output through speakers, these values are converted into sounds. Graphic data are similarly broken up into chunks, converted into scalar values, and arranged in a two-dimensional format that is output to a screen as an image.

"Meaning" refers to this value-added interpretation of information. In the computer example, it is particularly noteworthy that simply offsetting the starting point of the strings of ones and zeros (e.g. pointers in C programming) by one position renders the result totally meaningless. Furthermore, if you try to play image information as a sound, the result is simply noise. For the zeros and ones to possess any meaning, the software interpreting them must know how to split them up (in chunks of 32? of 64?), how to convert them into scalars, and what the resulting values are supposed to represent (audio encodings? image pixels?). In that sense, none of the ones and zeros in a computer has any meaning in and of itself. Indeed, in Shannon's theory, information is fully determined by the probability distribution on the set of such values, unrelated to their meaning.

Similarly, no single neuronal firing in the cerebral cortex carries meaning in the above sense. Regardless of whether a neuron fires in the visual cortex or in the auditory cortex, all it sends is an unremarkable set of action potentials. Not knowing the source of an action potential, it is impossible to tell whether it originated from an eye seeing a red apple, or an ear hearing the metallic tone of a trumpet. As a side note, subcortical neuronal firing prior to entering the cerebral cortex reflects differential signal characteristics stemming from various sensory organs, such as the eyes or ears. In the case of hearing, for example, neurons fire along with the phase of the heard sound. Interestingly, these characteristics are quickly lost upon entering

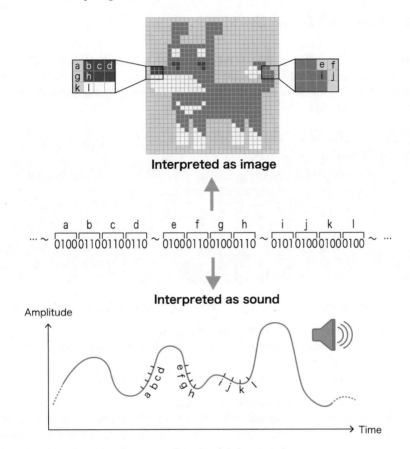

Fig. 5.3 Information only takes on meaning when it is interpreted

the cerebral cortex, where neurons basically emit action potentials randomly in time with modulating rates. Hence, regardless of sensory modality, cortical information is encoded in terms of which neurons fire and which do not. Thus action potentials in the cerebral cortex, where most scientists believe where consciousness resides, take on meaning only when it is processed by dedicated neural networks that are tuned to do so.

For this reason, I expect that the objective target of a natural law of consciousness is not neuronal firings themselves as information, but rather the neural algorithm that processes and interprets those firings.

Algorithms and Generative Models

Because some of my readers may be unfamiliar with what an algorithm is, allow me to introduce the oldest one known: the Euclidean algorithm. You probably learned this in a high-school math class as the way to find the greatest common divisor between two numbers.

Euclid's Algorithm

Given two natural numbers M and N $(M \geq N)$, repeatedly perform the following procedure to determine their greatest common divisor:

(a)   Divide M by N, and find the remainder R.
(b)   If there is no remainder $(R = 0)$, N is the greatest common divisor.
(c)   If there is a remainder $(R > 0)$, then find the greatest common divisor of N and R (replace the value of M with that of R, and go back to step a).

As a concrete example, let's apply Euclid's algorithm to the numbers 21 and 6:

1.   Dividing 21 by 6 leaves a remainder of 3 (step a).
2.   There is a remainder (step c), so next, we find the greatest common divisor of 6 and 3 (step a).
3.   There is no remainder when we divide 6 by 3 (step b), so the answer is 3.

As this example shows, an algorithm is simply a series of calculations performed according to the previous outcome. We perform one calculation, then use the results to perform the next. Following the steps until the algorithm completes yields an answer.

Similarly, a neural algorithm is a procedure for neural processing. Among the many neural algorithms that theoretical neuroscientists have proposed, my primary candidate for the objective target of a natural law of consciousness is an algorithm called the "generative model." To explain why, I will turn to Antti Revonsuo, a Finnish philosopher, who relates the neural mechanism of consciousness to virtual reality.

The Virtual Reality Within Our Brains

A dreaming brain is largely disassociated from the body's actual environment. The three-dimensional world of dreams is fully independent from the reality of lying in bed; it is wholly a fabrication of the brain. While dreams often have highly unrealistic features, the dream world recreates the real world with surprising accuracy in terms of adherence to physical laws. Gravity works, inertia works, and if you drop a plate, it will shatter into fragments. Reproducing realistic physics like this in the computer graphics of a Hollywood film may require months of computer time. Furthermore, you can shift your gaze, move your body, and feel your own weight in dreams. You often see other people and have conversations with them. While doing so, you do your best to gauge their intentions and motivations, despite the fact that your own brain wrote the script they are following. So our dreams reproduce not only the physical laws of a real-world environment, but also a body image that responds to dream-state motor commands, others with independent wills, and even our mental responses to them.

The movie *The Matrix* describes a world in which human brains are connected to an enormous computer that creates a highly realistic virtual reality, completely replacing sensory input and motor output. When we dream, it is almost like our brains are connected to the matrix system, creating their own inner virtual realities.

This raises an interesting question: Did the brain evolve and acquire such a sophisticated mechanism just for the sake of dreaming?

Antti Revonsuo replies "no." He argues that the inner virtual reality system is also operating during wakefulness. The very sensations you feel while reading this book, the whiteness of its paper and the smoothness of its cover, are all produced by the same neural mechanism at work during dreaming. In fact, it is the other way around: the brain acquired this system for use during wakefulness, and our nightly dreams are a by-product. Revonsuo calls this hypothesis the "virtual reality metaphor of consciousness," and he proposes that this inner virtual reality system is what generates our subjective experience.

If so, you might wonder, what is the difference between wakefulness and dreaming? During wakefulness, the brain's virtual reality system synchronizes with the external environment through sensory input and bodily feedback. Think of this as the system being anchored to the outside world. During sleep, the brain's virtual reality system loses this anchor, and therefore creates a world that is completely independent from the outside world.

a    **b**

Fig. 5.4 An example of change blindness (Reprinted from Ma et al. 2013)

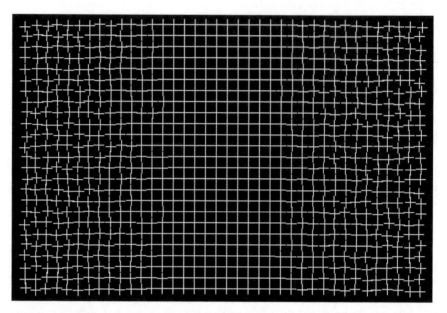

Fig. 5.5 The healing grid illusion (Reprinted from Otten et al. 2017, ©Otten et al. 2017)

Box 1: When the Brain's Virtual Reality Deviates from Actual Reality

Compare the two photographs in Fig. 5.4 and see if you can find any differences. While you are searching, note how until you find some point of dissimilarity, your brain is telling you that both photos are exactly the same. Next, stare at the center of Fig. 5.5. After a time, you should see the broken lines on the grid edges gradually repairing themselves. The paired photos demonstrate "change blindness," a phenomenon first academically defined by the American psychologist William James. The second illusion, developed by my close friend and collaborator, Ryota Kanai, is called a "healing grid."

Both of these illusions illustrate how your brain copes with a lack of visual information when creating the visual world you experience. There are just too many details in the photographs to hold in memory as your eyes travel back and forth between them, and this lack of information results in change blindness. Your brain tries to use context to fill in holes, for example, by assuming that a row of windows continues when in fact, it does not.

Something similar happens with the healing grid. When you stare at one spot for a long time, adaptation characteristics of your eyes cause you to start losing fine-grained visual information at the periphery of your field of view. Again, your brain tries to use context to fill in the lost information, assuming that since there is a neat grid pattern in the middle of the figure, it probably extends out toward the edges. An interesting aspect of this illusion is that if there are not at least fragmentary visual features at the periphery of vision (here, the broken lattices), this erroneous filling-in of fine-grained visual information does not occur.

As these examples show, while we often feel we are seeing the world in detail across our entire visual field, in truth, what we see does not always reflect what is present. Our senses do not directly monitor the outside world. Rather, the brain receives fragmentary information from sensory organs and uses this information to create the approximated virtual reality that we live in.

Brain's Virtual Reality During Wakefulness

Next, I will describe a phenomenon that points to the fact that the brain's virtual reality system is indeed functioning during wakefulness. The brain requires two components to maintain its virtual reality: an environment simulator that models the space around us, and a body simulator that models our own body. In a broad sense, the body is part of the environment from the brain's perspective. The phenomenon described below involves the brain's body-simulator.

You may have heard of "phantom limbs," a phenomenon in which amputees still feel as if their missing limb is present, and in many cases, feel that they can even move it (Fig. 5.6). In one case, a patient reported feeling as if he could reach out with his phantom limb to grab a mug placed in front of him. If he pulled too hard in an attempt to move it, he felt enough resistance to cause pain. This can only be explained by a body simulator in the brain operating during wakefulness, say, an arm simulator intercepting motor commands from the cortical motor area (coined the "efference copy"; copying the outward signal from brain to body), moving the simulated virtual arm accordingly, and returning simulated skin and joint sensations to the cortical sensory area.

The brain's body simulator mimics the body at such high levels that neuroscientist and neurologist Vilayanur Ramachandran found that we can subject it to physical rehabilitation. It has been known among medical doctors since the sixteenth century that if an arm or leg was paralyzed for a long time before amputation, the resulting phantom limb could inherit that paralysis. This would not be a huge problem if the symptoms ended there, but in some cases, the phantom hand has been clenched into a fist with phantom fingernails digging into a phantom palm, causing actual pain. Indeed, many amputees suffer from phantom limb pain of one type or another. But because no limb is present, doctors have limited treatment options. Some have attempted to alleviate pain through further amputation of the remaining stump, though any therapeutic effects were only temporary.

In an attempt to cure the real pains caused by phantom limbs, Ramachandran created a rather peculiar device comprising a box and a mirror (Fig. 5.7). Amputees

Fig. 5.6 Phantom limbs can
feel very real

Fig. 5.7 Ramachandran's
phantom limb rehabilitation

inserted their remaining intact arm into the box, and its reflection in the mirror made
it appear as if their amputated limb had returned. They were then told to open and
close both their actual and phantom hands at the same time. When they did so, visual
feedback from the mirror allowed the previously clenched phantom fist to seemingly
follow the brain's instructions, opening and closing as directed. Through several

weeks of continued rehabilitation, patients eventually learned, without aid of the device, how to manipulate previously paralyzed phantom limbs and finally relieve phantom pain.

These trials by Ramachandran demonstrate just how sophisticated the brain's body simulator is, and hopefully have convinced you that it is undoubtedly functioning during wakefulness and further generating our bodily sensations. In the next section, we return to where we left off, the generative model, to see how this virtual reality system can be implemented as a neural algorithm.

Generative Model as a Neural Implementation of Brain's Inner Virtual Reality

Sensory processing in the brain was once thought to be realized through a series of processes that move from lower- to higher-level cortical areas in a bottom–up manner. Indeed, Chap. 2 of this book described visual processing in terms of this classical view. However, it was overturned by the generative model, which was independently proposed by Mitsuo Kawato and Toshiro Inui, and David Mumford in the early 1990s.

Generative models reverse the direction, treating top–down processing from higher- to lower-level streams as the main process. The "generative" part of a generative model refers to its generation of estimates of lower-level neural activity based on higher-level activity. Generated estimates are compared with lower-level activity, and the error between them is calculated. This generative error is then used to update higher-level activity, so that it better matches the external world. This process iteratively repeats until the error converges to a minimal value.

The generative model's objective is to ensure that activity in higher-level visual areas properly reflects the external world, so regarding this point, it has the same end goal as do conventional models. The difference is that calculations of higher-level neural representations are not performed in one shot, but rather verified through a generative process (For further details, see Appendix).

Towards Realistic Virtual Reality

You might be wondering why our subjective experiences seem so real and three-dimensional if their source is a generative process whose sole objective is to estimate lower-level neural activity.

Consider a generative model that can flexibly respond to any situation. To do so, we must supply the higher-level visual area with a large variety of symbolic information as the source of the generative process. First, we need neurons that code the existence of various visual objects such as "houses" and "trees." We also need neurons that code the properties of those objects, such as their three-dimensional structure and how their surfaces reflect and absorb light. Next, we need neurons that determine the three-dimensional configuration of multiple objects. For example, for every object in the visual field, we need information describing that object's position and orientation. Finally, we need neural representations for various light sources.

Taking this higher-level symbolic information as input, the goal of the generative process is to estimate lower-level neural firings driven by sensory input. So what kind of generative process would lead to precise estimation in a variety of visual configurations and scenarios?

As we move around in the three-dimensional world, we change the relative positions between ourselves and the visual objects within it. For instance, let's assume that there is a tree in front of us, and a house behind the tree. If we move to the right, both objects move to the left, but the tree in the foreground moves farther. In contrast, if we move toward the center point between two objects, the object on the left moves left, and the object on the right moves right. In the end, it seems that the easiest way to generally estimate neural firings in low-level visual areas is for the generative process to simulate the three-dimensional world as-is.

Consider how computer graphics are created. For example, you might want to create a scene that shows a house with a tree in front of it, illuminated by afternoon sunlight (Fig. 5.8). To do so, you would create three-dimensional models with the desired shapes, then apply appropriate textures to their surfaces. After placing these objects in the scene, you would simulate how light travels and reflects within it. Light emitted from sources will reflect off objects in the scene, and the characteristics of this reflected light must be calculated according to the light absorption and reflection characteristics of object surfaces. After all of these procedures, you will have established a well-lit three-dimensional virtual world. The final step in computer graphics is to place a virtual camera within that world. Doing so determines the observer's position and field of view, and the captured 2D scene becomes the final image.

Applying something like this "computer graphics" strategy in the generative process would make it extremely flexible in accurately reproducing low-level neural representation. The brain would acquire its own internal virtual world, one that follows the same physical laws as the external world and can mimic any situation. Notably, such realistic generative models would serve as a perfect neural implementation of Revonsuo's hypothesis, and explain the rich and vivid visual world that we subjectively experience during wakefulness and dreaming.

Box 2: Reconsidering Natural Laws of Consciousness

If there are natural laws of consciousness, they have likely existed from the moment of the Big Bang; given the very nature of natural laws, it is hard to imagine them arising at some point in the vastness of space and evolving along with the universe. It is therefore unlikely that they apply only to central nervous systems like those found on Earth. So then, what would a generalized version

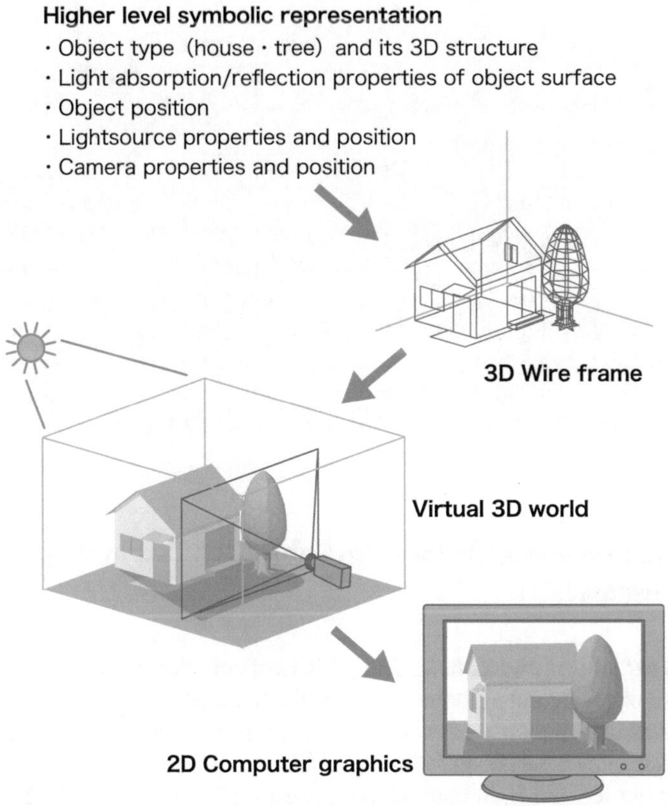

Higher level symbolic representation
- Object type (house · tree) and its 3D structure
- Light absorption/reflection properties of object surface
- Object position
- Lightsource properties and position
- Camera properties and position

3D Wire frame

Virtual 3D world

2D Computer graphics

Fig. 5.8 Creation of a computer graphics image

of the neural algorithm hypothesis look like? We can think of an algorithm that functions as a mirror that reflects the world. This mirror, however, reflects not only surface appearances, but also causal relations and the underlying physics. Perhaps conscious sensory experiences arise during this reflection process, with varied intensity and richness depending on the capacity of the processor. Humans have highly refined senses of vision, hearing, and touch, and thus have vibrant conscious sensory experiences arising from these senses. Dogs, in contrast, likely have much more vivid conscious experiences regarding olfaction. We can imagine them as "seeing" colored particles arising from odor sources.

Moreover, reflections in the generative model do not come solely from what we normally think of as the outside world. Sensory experiences such as body and organ sensations also arise during this reflective process, since from the brain's perspective, our muscles, heart, and other body parts are external. Taking this concept even further, we can posit that, from the perspective of a conscious brain, even some brain functions are part of the reflected outside world. For instance, decision-making that requires snap judgments—like those performed by our baseball player in Chap. 3—occur in some place inaccessible to the consciousness mechanism. The illusion of conscious free will itself may therefore be another reflection arising from consciousness mechanisms.

Getting back to the original discussion, I believe that a general form of the proposed neural algorithm hypothesis will be based on these "reflections." In other words, when something reflects causal relations of some other matter, subjective experiences for the reflected matter arise, even when this process does not occur within an earthly central nervous system. If that is the case, then perhaps consciousness has indeed already arisen in, say, self-driving automobiles.

Will Generative Models Pass the Subjective Test for Machine Consciousness?

As we did earlier for two candidate natural laws of consciousness, let's speculate on the validity of the neural algorithm hypothesis. In particular, as with the treatment of integrated information theory, we will examine whether the generative model provides some key mechanism for unifying the two streams of visual consciousness situated in our left and right cortical hemispheres. In other words whether our two cortical hemispheres—which produce a unified subjective experience despite their restrictions in interhemispheric connectivity—can be viewed as a single generative model.

First important element of the generative model is the symbolic representation in its higher levels. In the computer graphics example in Fig. 5.8, these symbolic representations must be shared between the right and left cortical hemispheres in order to accommodate visual objects that span the right and left fields of view. Furthermore, the effects of a light source on one side would obviously extend to the other, so again, sharing of such symbolic information is crucial.

Conveniently for this sharing of symbolic information, as we saw in the previous chapter (Fig. 4.14), higher-level visual areas such as IT possess interhemispheric connections across a wide range of the visual field. Such connectivity that bridges neurons manifests as large receptive fields that span both visual fields, which can be regarded as experimental evidence for bilateral sharing of high-level symbolic information. We can thus conclude that high-level visual areas satisfy the requirement of operation as a single generative model.

The other essential element of the generative model, the generative process, is a little more complicated. In this case, the low- and mid-level visual areas that are responsible for this process (the primary through fourth visual areas) are basically isolated; interhemispheric connections only exist between cortical subregions corresponding to the border between the left and right visual fields (the "vertical meridian" in Fig. 4.11).

So the question is whether this semi-isolation affects the generative process. Let's use the example in Fig. 5.8 to trace through each step. To begin with, the first conversion from high-level symbolic representations to three-dimensional structures and surface textures is minimally affected. This is because, so long as high-level symbolic representations are shared, structures and surfaces in both fields can be independently defined.

Given that the three-dimensional virtual world is now structured and surfaced, the next processing steps are to light it and obtain a virtual camera image. We must calculate how light emitted from specific sources hits surfaces, then reflects toward the virtual camera according to the reflection characteristics of each surface. After doing so, light from all surfaces arriving at the virtual camera—including that arriving directly from the light source—will form an image on its virtual sensor, completing the generative process.

Here again, there is basically no crossover of processing between the right and left fields of view. Information related to light sources (the type of light, its direction and position, etc.) can be directly derived from shared symbolic representations, and its interaction with object surfaces can be calculated basically independently. Why do I say "basically," here and above? Because there could be cases in which, for example, a mirror in one field of view reflects some object in the other, creating the need for an interhemispheric generative process. Even in such cases, however, if high-level symbolic information can be extended to include a "projection of reflected objects" for mirrors and such, then it can be reproduced without right–left crossover. Therefore, we may conclude that all stages of the generative process are confined within their respective hemisphere, and thus, require no interhemispheric connectivity.

In summary, limited interhemispheric connectivity in low-to-mid-level visual areas does not constrain the generative process. In light of the previous discussion, where we showed that widespread high-level interhemispheric connectivity assures sharing of high-level symbolic information, we may now safely conclude that a single generative model spans the two cortical hemispheres, despite the biological constraints of interhemispheric connectivity (Fig. 5.9). So coming back to our initial inquiry, the generative model indeed provides a key mechanism in unifying the two streams of visual consciousness.

As an aside, we currently lack direct experimental evidence that the biological brain simulates a three-dimensional virtual world. Yes, our three-dimensional visual experiences, especially during dreaming, strongly suggest this is the case, but as of today, neuronal measurements have not conclusively demonstrated it. One reason for this is the use of two-dimensional visual displays in mainstream macaque studies, limiting the presentation of three-dimensional scenes. Another reason might be the high complexity of neuronal response properties in the areas responsible for the generative process (see Box 2 in Appendix), where such representations may be indecipherable with current analytical tools. Although, I am quite certain that future

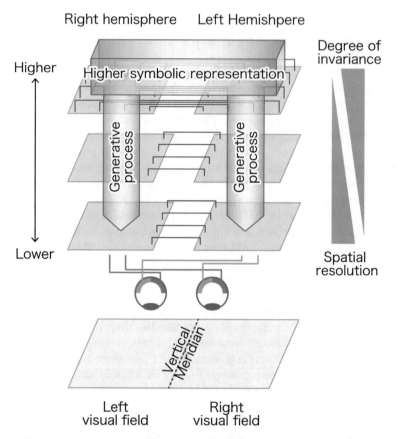

Fig. 5.9 A generative model that produces integrated sensory experiences without being impacted by the anatomical and physiological constraints of a brain divided into right and left hemispheres

developments in stimulus presentation and data analysis will shine new light on this subject.

Box 3: Given That Generalization Increases in Higher-Level Visual Areas, How are High-Resolution Generative Processes Possible?
In reading the discussion above, you might have wondered whether any high-resolution visual information exists in higher-level visual areas. And if not, wondered how it can generate a high-resolution 3D virtual reality world that matches our subjective experience. This is an important issue, and I believe there are two possibilities.

The first possibility is that the above problem of irreversible (lossy) compression in high-level visual areas is solved through a multi-layer generative model. By multi-layer, I mean that multiple submodules for generative processes exist, and these submodules are stacked one on top of the other as elegantly formulated by Rao and Ballard in 1999. The lower-level areas in one submodule would play the role of higher-level areas in the submodule below, with "representation of symbolic information" and "representation of sensory input information" relatively defined within the context of each submodule. This type of modular construction would allow the imperfection of symbolic representations at the highest level to be absorbed through generative errors at various levels. The generalization feature of neurons described in Chap. 2 sheds some light on this idea. The sloppiness in the high-level symbolic representations can be gradually refined by blending it with residual generative error via high-resolution sensory input, thereby producing a photorealistic virtual three-dimensional world. If this is the case, then the visual experiences in our dreams should be far less detailed than what we experience when awake, lacking sensory input and the consequent hierarchical error absorption process. Indeed, I can recall dreams in which I read relatively large writing on a piece of paper, but I cannot remember any in which I read the small print of a newspaper article. Can you?

The second possibility is much more controversial, but even so, I believe it requires serious consideration. This is the possibility that, contrary to well-established views in neuroscience, high-resolution visual information is in fact preserved in high-level visual areas. In other words, there is no irreversible compression, but only transformation of information. If this is the case, since individual neurons in high-level areas have large receptive fields and do not divide the visual world into small segments, information would be represented in some very complicated format that could not be deciphered through measurement of only a few dozen or even a few hundred neurons. Interestingly, the laboratory of Yukiyasu Kamitani, whom I mentioned in Chap. 3, has produced experimental results that may suggest just such a scenario, by applying an innovative fMRI methodology ("multivariate analysis") that Kamitani himself went on to develop after his Shimojo-lab years.

However, even if we suppose that high-resolution visual information is preserved in high-level visual areas in some abstract form, that alone is likely insufficient to establish our visual subjective experiences; it requires at least mid-level visual areas. As evidence, the British neuroscientist Semir Zeki reported a case in which localized damage to the fourth visual cortex, a mid-level visual area in the ventral pathway, resulted in the patient becoming unable to visually dream.

Explaining Time Lags in Consciousness with the Generative Model

Chapter 3 described how conscious time lags hundreds of milliseconds behind physical time. A generative model would explain this lag as follows.

One feature of the generative model is that a series of visual processes are iteratively computed until the generative error is minimized. Generative error is gradually reduced as these processes repeat, and as a result, our inner virtual world becomes more and more synchronized to the real world. This implies that, at early transient stages before generative error is minimized, the inner virtual world will unavoidably be full of contradictions and mistaken identities. If our conscious vision had full access to this transient period, viewing a monkey's face, for example, would result in seeing a "monkey–human" hybrid that gradually morphs into an increasingly monkey-like form. In fact, such gradual changes in the coded information of neurons have been demonstrated in a paper by Sugase and her colleagues in 1999.

Of course, we never actually see such strange beasts. Assuming the generative model yields our subjective experience as a neural algorithm, solving the problem of keeping intermediate transient neural activity out of our conscious vision is straightforward. All we need in the neural algorithm is a rule that says, "generate subjective experiences only after the generative error has converged to a minimal value." Interestingly, that rule will unavoidably delay sensory input entering our conscious vision, in line with Libet's findings. We may postulate that our visual system wants to show us only a sensible world, so much so that it is willing to make us wait until it has cooked one up.

On the other side of the coin, subconscious high-speed processing of visual information like that described in Chap. 3 can be also explained using the generative model with some help again from our baseball player. Circumstances do not allow for delay when up at bat, so above all else, visual processing needs to be fast. In such cases, it makes perfect sense for the generative model to provide downstream decision-making areas with information that underwent only the first pass of iteration. This first pass visual information would be provided immediately, but because generative error has not yet been minimized, it is not allowed to rise to consciousness. Subjective experience comes only later, after a sensible world has been formulated in the final pass.

This relationship between conscious and subconscious processing in the generative model also provides an appealing explanation to the inner workings of Libet's subjective back-referral. Since the raw information of sensory input including its timing is there, it is not much of a problem to link the perceived timing to it, once our subjective experience is established. We may even go one step further and apply the same principle to our decision-making process. In that case, Libet's assertions that we do not possess conscious free will and that the actual decision-making process is only occurring subconsciously, can be described by the subconscious first pass of the generative model. Meanwhile, the succeeding conscious confabulation can be attributed to its final pass.

Lastly, putting the generative model aside, the basic idea of relating subconsciousness to the first pass of bottom-up cortical processing and consciousness to recurrent cortical processing was initially proposed by Lamme and Roelfsema in their influential paper published in 2000. The two came up with their concept upon careful examination of their own neural recordings from the macaque visual system. Future work in a similar direction of investigating the inner workings of the biological generative model would be of great interest.

The Generative Model Explains Differential Subjective Experience Arising from Various Sensory Modalities

Taking the generative model as the source of consciousness has further benefits. For example, you might have wondered how some neuronal firings produce visual experiences, while others produce auditory experiences. The short answer can be found by returning to the discussion of the meaning of information in the first half of this chapter, but let's look a little more closely at the matter.

Solving this question seems challenging if we assume that the source of consciousness is neural information of any sort. Chalmers' theory has nothing to say about differentiating types of information, but integrated information theory would also likely suffer. Their measure for integrated information differentiates what enters consciousness from what does not, but not the types of multimodal information that leads to very distinct subjective experiences.

In contrast, posing a neural algorithm as the source of consciousness—and specifically the generative model—solves this problem for free: differential statistical properties of various sensory modalities are strongly reflected in the generative process. In the case of vision, for example, a line segment located at some point in space is very likely to continue in its vicinity, and it is reflected in the generative process. In the case of hearing, there is a high possibility that a sound wave heard at some point in time is contiguous with immediately preceding and subsequent time points, and this too must be reflected in the generative processes.

If the generative process did not reflect these natural statistics of sensory signals, it would not achieve its primary goal in the first place: using high-level symbolic representations to accurately reproduce neural firings in low-level sensory areas. The objective goal of the model assures the acquisition of differential generative processes for various sensory modalities, so it naturally explains qualitatively differing subjective experiences.

Another interesting issue concerns the information that rises to consciousness. As I described in Chap. 2, lower-level visual areas in particular overflow with information that never reaches consciousness. Examples in the primary visual cortex include raw visual information before corrections for color constancy, and before corrections for rapid involuntary eye movements.

If we consider a neural algorithm to be the source of consciousness, selecting which information rises to consciousness is simply a matter for the algorithm to decide. One example would be that the top-down generative process enters consciousness, but generative errors and bottom-up sensory information do not. It would not be troubled by, as we saw in Chap. 2, the highly-intermingled nature of neurons representing conscious and subconscious information throughout the cortical hierarchy. In fact, it would even explain why they are intermingled; generative processes sit side by side with the processing of generative error and bottom-up sensory input in multi-layer generative models (see Box 3).

Box 4: The Web of Causality Through Neural Algorithms and Deterministic Chaos

---

Fig. 5.10 Integration of fluctuations in neural activity and a neural algorithm through deterministic chaos

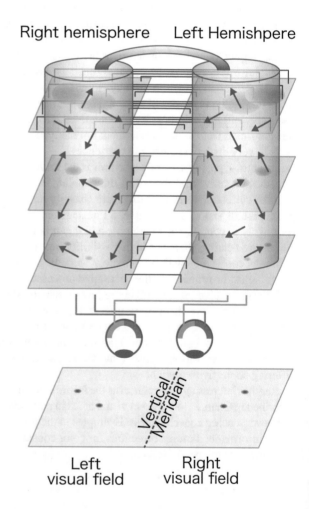

Box 2 in Chap. 4 discussed causality in neural networks. It is also interesting to consider causal relations through the concept of deterministic chaos.

Deterministic chaos describes situations where there are no random factors, but nonlinearity makes it impossible to predict future states from observation. Simply put, nonlinearity occurs when one plus one does not necessarily add up to two, so as a matter of course, the thresholding property of neurons is nonlinear. Thus, it is considered likely that deterministic chaos arises amidst the complex interweaving of these nonlinear elements in the brain.

Deterministic chaos leads to interesting effects. Much like the so-called "butterfly effect" (a butterfly flapping its wings in Brazil sets off a tornado in Texas), the slightest perturbation in a neural network can linger, rather than eventually fading away, and continue to grow until it spreads out to the entire system. This can result in a state in which all neurons in the neural network are enmeshed in a "web of causality" (Fig. 5.10).

Current measurement technologies do not allow us to capture a full picture of deterministic chaos within a brain, but many reported observations suggest its existence. For example, the timing of action potentials from any single neuron has no regularity, but rather varies apparently randomly around a modulating firing rate.

In reality, because quantum effects are at work in the spatial scale of opening and closing of ion channels, the brain is constantly rolling dice at miniature scales. Thus, the brain's operating point is likely a mixture of deterministic chaos and stochastic noise, but nonetheless, the web of causality should be there, but with a shorter lifespan.

What I am trying to postulate here is the possibility that semi-deterministic chaos functions as a kind of a "glue," through its widespread web of causality, to bind neurons as a neural algorithm within and across cortical hemispheres. The very early version of my ideas regarding the test for machine consciousness, focused more on this aspect, as you can see in a video of my 2014 talk given at the Redwood Center for Theoretical Neuroscience at UC Berkeley. In relation, it is particularly interesting that Tononi and Gerald Edelman's "dynamic core hypothesis," a forerunner of integrated information theory, attributed similar roles to deterministic chaos.

Dual Aspect Theory of Generative Process

Up to this point, I have primarily focused on vision as an example modality when discussing generative models and consciousness. However, the brain's virtual reality consists of many modalities other than vision, and subjective experience as a whole likely arises through numerous generative models distributed across the cerebral cortex. If that is the case, generative models should simulate and estimate inputs not only from bodily receptors that reflect the outer world, our body and internal organs, but also from subcortical neural systems within the brain. One such example would be the amygdala, a subcortical brain region responsible for physical fear responses (e.g., dilation of the pupil, sweating, increased heart rate), where the subjective experience of fear would presumably emerge from a cortical generative model that simulates the amygdala-body interaction as a whole. Once we manage to validate the natural law of consciousness using vision, where we have in hand the proposed subjective test for machine consciousness to pull it off, my expectations are that other modalities would soon follow by making use of scientific knowledge that we gain there.

In closing, I offer a Chalmers-esque definition of dual aspect theory of generative process: Generative process has both the objective aspect of performing information processing, and the subjective aspect of yielding subjective experiences.

Chapter 6
Uploading Human Consciousness

Viable Method for Mind Uploading

Many scientists question the very purpose of machine consciousness. If our only goal is to develop a "conscious-like" machine, the authenticity of that consciousness is unimportant. As long as the machine *behaves* as if it is conscious, we don't need to determine whether it genuinely is. If our goal is to upload our own consciousness, however, things are very different; we certainly wouldn't want to find ourselves transformed into a philosophical zombie after the process.

But could human consciousness truly reside in a machine? Ray Kurzweil, Director of Engineering at Google, predicts that this will be possible by the middle of this century. If so, what would the process of mind uploading look like? In film and television depictions, consciousness transplantation usually occurs through some noninvasive procedure with some fancy device scanning the brain from outside the cranium in a matter of minutes. Having read this far, you no doubt see how this would be impossible; transplanting human consciousness into a machine will be an invasive time-consuming process.

In this final chapter, building on top of the proposed subjective test for machine consciousness, I introduce a transplantation process that consists of three stages—constructing a machine with "neutral" consciousness, connecting and integrating our own consciousness with it, and finally, transferring our memory to the machine. Notably, unlike other proposed methods that attempt to make digital copies of our brains, we wouldn't need to die first and taken our brains out of the skull. It would be a totally seamless procedure.

Prospects for Machine Consciousness

Making predictions about which candidate natural laws of consciousness we should first implement in our machine is tricky business. If we subscribe to Chalmers' dual

© Springer Nature Switzerland AG 2022
M. Watanabe, *From Biological to Artificial Consciousness*, The Frontiers Collection,
https://doi.org/10.1007/978-3-030-91138-6_6

aspect theory of information, then conscious machines should already exist. After all, if moon rocks possess consciousness, then surely we can assume the same for digital cameras.

Of course, most would say the likelihood of that being the case is low. A surer approach to our quest for machine consciousness is to mimic the human brain, which we know for certain that it possesses consciousness. And doing so would provide an added bonus: if our machine ever yields consciousness, it will more likely be compatible with our own, and thus, more likely to integrate. The validation of natural laws of consciousness can come later. We may take full advantage of the machine's degree of freedom and conduct manipulation experiments that are unimaginable in the biological brain, and consequently, pinpoint exactly which candidate natural laws need to be minimally met to maintain its consciousness.

If Chalmers's take on the fading qualia thought experiment is correct, then massively parallel computer hardware that reproduces the brain's network structures will become conscious. But will this ever be possible?

In fact, researchers have been developing "neuromorphic chips," or semiconductors equipped with artificial neural networks, since the 1980s. In the summer of 2014, IBM announced TrueNorth, a neuromorphic chip with one million spiking neurons and two hundred million synapses, and this chip was featured on the cover of *Science* (Fig. 6.1). One million neurons are on the scale of the central nervous system of a

Fig. 6.1 *Science* cover featuring IBM's neuromorphic chip TrueNorth

cockroach, and within sight of those in the mice that I use in my daily research, which have around seventy million neurons.

However, to some extent, this is an apples-to-oranges comparison. TrueNorth has 4096 physical CPU cores, each of which has 256 neurons that must be processed through a time-sharing system. This means only 4096 neurons will be active at any time, far fewer than the number of neurons in the entire system. Another factor is the precision with which these artificial neurons mimic biological ones. Neuromorphic chips are still in early development, and the trend is to simplify their neurons to the extent possible. Their current precision is highly unlikely to fulfill the requirements of fading qualia, namely replacing biological neurons without being noticed.

Not that this is unexpected. Neuromorphic chips are designed to allow the implementation of large-scale neural networks at low power consumption, not necessarily to harbor consciousness. If we want chips suited to the latter task, scientists will need to produce persuasive evidence and theories demonstrating their viability.

To that end, I have cast my lot with demonstrations of digital fading qualia to show a reasonable possibility that consciousness can reside in a neural network simulated on a von Neumann computer. Thankfully, many specialists are optimistic that computers capable of simulating the entire human brain in real-time will be available by 2030. In such simulations, the dynamics of a neuron would be modeled using hundreds of small segments, each calculating a dedicated set of differential equations. Such a system, the so-called multicompartment neuron model, would likely satisfy the requirements of the digital fading qualia thought experiment.

Complications in Perfecting a Digital Copy of Our Brains

While the original fading qualia and digital fading qualia thought experiments suggest that consciousness can reside in a machine, its end product, a perfect copy of our brains, cannot be physically realized. The process of replacing a biological neuron with an artificial one, or of replicating its functioning in a computer, results in an end product that replicates all of its former synaptic connectivity. But this is simply not possible.

To replicate biological synaptic connectivity, we would first need to quantitatively decipher its values in a living brain. A scientific method for doing so exists, but it requires opening the skull and attaching to the brain surface a special kind of microscope (a two-photon microscope) that can measure ionic flows at the dendrite. Furthermore, recording a single synapse with sufficient precision requires time in the order of seconds. Even considering future technical innovation, it is difficult to conceive of any advance that would allow measuring the 100–1000 trillion synaptic connections of an entire human brain at practical times. 100 trillion seconds amount to 3 million years, and speeding up multiple orders of magnitude would not be very helpful.

What about then acquiring synaptic connectivity from postmortem brains? In fact, there has been remarkable progress in the field in recent years, where reading the full connectivity of increasingly large brains has been achieved, starting from the central nervous system of *C. elegans*, which has only 302 neurons, to that in fruit flies with approximately 100,000 neurons.

The basic strategy is to slice the brain at thicknesses on the order of tens of micrometers, then apply scanning electron microscopy to track the axons in three dimensions. The problem is, while extracting binary connectivity is quite straight-forward, acquiring quantitative synaptic efficacy remains a challenge. As we saw in Chap. 1, in order to obtain quantitative values, we need to observe ion channels at the synaptic cleft at much smaller spatial scales, on the order of nanometers. There are methods for visualizing clusters of specific types of ion channels that allow quantization to some extent, but acquiring the full connectivity of a human brain with sufficient resolution that leads to perfect reproduction of the original neural dynamics seems unrealistic.

But for the sake of argument, let's assume that this does become possible at some point in the future. Would you risk uploading yourself by such a process? The deal-breaker for most is that this method requires slicing up a postmortem brain, so uploading could only occur after we have died and our brains are extracted from our skulls. Not that doing so would be completely without merit; you might find some solace in knowing that "something" of yourself would carry on. But personally, I would prefer a method that provides a little more continuity.

Constructing First a Neutrally Conscious Machine

Instead, as I mentioned at the beginning of this chapter, my plan is to prearrange a machine with neutral consciousness, connect it to our own brains while we are alive, and gradually "color" it with our own consciousness, in other words, transferring our memory. Here, we focus on the first step.

The basic strategy is to prepare a spiking neural network that replicates the full binary connectivity of the human brain and use it as an initial state for subsequent training. For binary connectivity, we can turn to scanning electron microscopy and acquire it from human brains, as described in the previous section. From there, for instance, to develop it into a visual system, we can show it a life's worth of video material. If we find that the network needs a body that interacts with its environment, we can supply it with a virtual body.

Critically, to apply such a training process, we need to specify the general system architecture. This is where the proposed neural algorithm hypothesis comes in handy; assuming that the visual system is a generative model, we can define an appropriate loss function as in modern deep learning. Another advantage of the generative model architecture is that it is an autoencoder; the output reconstructs the input. Due to this basic structure, we may apply error backpropagation type of learning, a key training method for deep learning (see Appendix for its descriptions). Here, error backpropagation requires teaching signals that specify the desired output of each and every neuron in the output layer, which the brain does not always have access to. In the case of a visual generative model, retinal information would serve as both the input and the teaching signal. Regarding the availability of error backpropagation like methods for spiking neural networks, a biologically plausible method has been proposed by Payeur and others in 2021. By making full use of advanced versions of the above methodologies, we are not too far from constructing a machine that is worth connecting to our own brains, something that we can be reasonably sure that the natural law of consciousness has kicked in.

Prospects for an Invasive Brain–Machine Interface

To prepare for the second stage of the uploading procedure, connecting the machine to our own brains, we need one more ingredient; a brain-machine interface(BMI) capable of connecting machines and brains at the neuronal level. Regardless of whether the source of consciousness lies in information or in neural algorithm, integration of machines and brains will definitely require sending and receiving signals via action potentials. This cannot be achieved through non-invasive methods like brainwave measurements, fMRI, or TMS, and so demands an invasive interface.

In 2017, the U.S. Defense Advanced Research Projects Agency (DARPA) launched its Neural Engineering System Design program, a four-year initiative with the goal of simultaneously recording and stimulating one million neurons. One group is investigating the use of ultra-small devices, smaller than a grain of salt, which will be embedded directly within cortical gray matter (Fig. 6.2 top right). This approach will require breakthroughs in device miniaturization, long-lasting power supplies, and wireless hyper-parallel communication, but if it becomes practical, it will be an ideal solution to some challenges related to BMIs that I present below. Meanwhile, another group is investigating the insertion of ultra-thin wire bundles, around one-tenth the thickness of a human hair (Fig. 6.2 left and bottom right). This approach has been used in many previous experiments, if not at this massive scale. I myself have used microwire bundles in mice, and believe they would provide a fine way of constructing a long lasting BMI.

The challenge in long-term BMI is recording and stimulating the same set of neurons over days, months, or preferably years. The first requirement for accomplishing this is to ensure that electrodes do not move around within the brain, since the slightest movement would make it impossible to track the same set of neurons over time. The movement would also cause wires to damage neural tissue, and should an electrode pierce a blood vessel, the resulting bleed could kill neurons, or cause irregular firings due to resulting changes in the characteristics of the surrounding cerebrospinal fluid. Here, the challenge in preventing electrode movement is the brain's tofu-like consistency. With the neurograin approach as one of the rare exceptions, electrode probes have to be anchored to the skull, so their relative positions will change any time the brain moves within the skull.

A second requirement for maintaining an interface with a fixed set of neurons is preserving the recording quality over long periods in the face of immune responses to

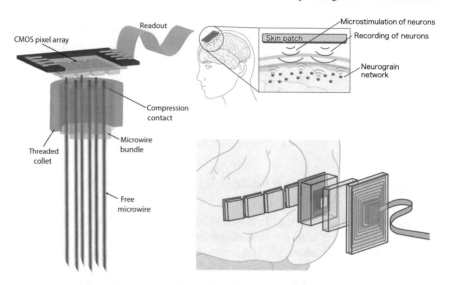

Fig. 6.2 A DARPA project aiming at an invasive BMI for one million neurons. The top right image shows the "neurograin" network approach by Arto Nurmikko at Brown University, and the left and the lower right image show a microwire electrode approach by Paradromics, Inc. (left: Reprinted from Obaid et al. 2020)

foreign substances. Tissue adhesion to electrodes alters their electrical characteristics and blocks the recording of action potentials. The larger the surface that surrounds the electrodes, the higher the probability that tissue adhesion will occur.

The advantage of microwire electrodes is that they minimize both of the above complications. They are so thin and flexible that they can deform along with brain movement, suppressing any unwanted relative motion. Further, since the cut tip of an insulated microwire functions as an electrode, its minimally small surrounding area makes tissue adhesion less likely. In fact, some groups have successfully applied microwire bundles to record action potentials from the exact same set of neurons for over a year.

A similar approach, also adopted by Neuralink, is to use flexible polymer probes with multichannel electrodes. The advantage over microwire bundles is that more neurons can be recorded along the cortical depth, due to the vertically embedded multiple electrodes along the probe (Fig. 6.3). However, these probes are less flexible than microwires, and their flat probe design may result in more tissue adhesion. More work is needed to determine the better approach, microwire bundles or polymer probes, in terms of long-term stability.

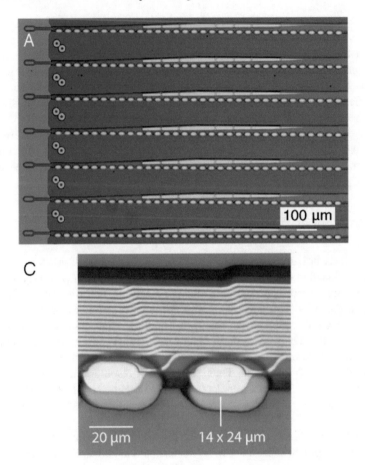

Fig. 6.3 Polymer probe electrodes by Neuralink (Reprinted from Musk and Neuralink 2019)

Possibilities for Machine–Brain Hemispheric Integration

Suppose DARPA's plan succeeds, and we can simultaneously record and stimulate one million neurons. Through extensions of this technique, how likely is it that we will be able to integrate our consciousness with a machine's? Below, we will take a closer look at the human brain in an attempt to derive a lower boundary on BMI capacity.

In Chap. 4, I described corpus callosotomy and the works of Sperry, but there is more to this story. Michael Gazzaniga, a former pupil of Sperry and currently a professor at the University of California, Santa Barbara, attempted to alleviate the serious side effects resulting from this surgery. He reported that as long as one of the three connective nerve bundles, the anterior commissure, was left intact, conscious vision remained unified and other side effects were largely prevented. As Fig. 4.12

shows, the anterior commissure has far fewer nerve fibers than does the corpus callosum—only twenty to thirty million in the human brain.

Interestingly, the anterior commissure is known to specifically connect high-level brain areas. This supports what I discussed in the previous chapter, that a shared high-level symbolic representation might be sufficient to unify the two cortical hemispheres as a single generative model, given that it generates consciousness.

So there is a good chance that a BMI with a capacity equivalent to that of the anterior commissure is sufficient to minimally integrate biological and machine consciousnesses. If that is the case, then once the DARPA project has attained its goals, we will have realized approximately one-twentieth of the capacity needed.

The Problem of Writing Information

However, there is a fundamental problem in *writing* high-resolution information with electrodes placed in the cortical gray matter. To begin with, doing so requires minimizing the number of neurons stimulated per electrode, because different neurons, even adjacent ones, can have distinctive stimulus preferences. If the passing of current through a single electrode activates multiple neurons, written information will correspond to a blurred average of multiple stimulus preferences. For example, if "dog" and "cat" neurons were simultaneously activated, it would be equivalent to writing "dog–cat" information. Thus, writing high-resolution information requires minimizing activated neurons per electrode, but as reported by the group led by Clay Reid, who is now a senior investigator at the Allen Institute, this is where the problem arises.

His group combined two methods for their experimentation. The first is two-photon microscopy, which allows two-dimensional imaging of thousands of neurons in cortical gray matter spanning a few hundred micrometers. The second method is electrical stimulation with a standard single-channel electrode. By observing exactly which neurons activate during electrical stimulation, the group found that even the smallest currents resulted in widespread activations. For instance, when they lowered the current to below 10 μA, resulting in only a countable number of neurons being activated within the scope of two-photon microscopy, neurons hundreds of micrometers away (Fig. 6.4a, c), or even 4 mm away (Fig. 6.4e), were activated. After conducting additional experiments to resolve the underlying mechanism, they concluded that the firing of distant neurons was due to their axons extending near the electrode tip. When current passed through the electrode, those nearby axons emitted action potentials, which traveled backward to their somas and triggered the firing.

The firing of distant neurons becomes a critical problem due to the limited distance at which we can observe action potentials, approximately 100 μm from the electrode tip. So even if we applied some minimal current value, there would always be neurons firing beyond the scope of observation. Without observing those firings,

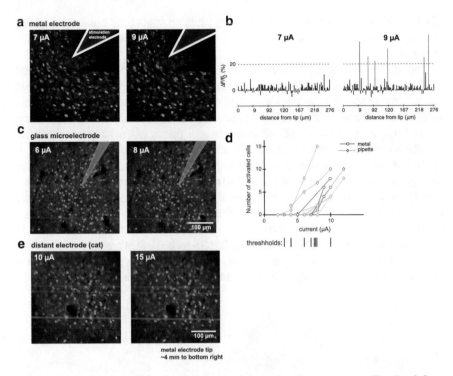

Fig. 6.4 Restrictions in writing in high-density information in gray matter (Reprinted from publication Histed et al. 2009, Copyright 2009, with permission from Elsevier)

there is no way to analyze their stimulus preference, and thus grasp what information is actually being written by passing current through an electrode. It would be like shooting in the dark. This mismatch in the spatial extent of reading and writing information complicates the construction of high-resolution BMI, as Reid and colleagues concluded in their paper: "Because a single site in the cortex activates neurons that are spread widely from that site, achieving high-resolution rasterized visual percepts by electrical stimulation through high-density electrode arrays may not be possible, unless the brain can learn to interpret these distributed patterns." This problem becomes even more critical if we wish to connect an artificial spiking neural network to our brain, for instance, to substitute functions of lesioned brain areas, not to mention unifying our consciousness with that of a machine.

A New Type of BMI: Insertion of 2D Electrode Array into a Dissected Axonal Bundle

I certainly hope that conventional BMI methods will overcome the above issues with some technical innovation, but as a safety measure, I would like to propose a new type of BMI that solves the read–write mismatch problem, and at the same time, allows full access to all neurons accountable for integrating the two biological hemispheres (Fig. 6.5).

The idea is to split nerve fiber bundles such as the corpus callosum and insert between them a double-sided, two-dimensional electrode array. This configuration would provide access to axons where they are straightly aligned, solving the read–write mismatch problem; recordings of action potentials and electrical stimulation would both occur at the tip of individual axonal fibers, where they are closest to specific electrodes within the array. Moreover, this is precisely where transhemispheric axons are concentrated, maximizing BMI capacity while minimizing tissue damage. In contrast, even inserting a 15-μm microwire into gray matter, the least invasive conventional method, is highly damaging to neural tissue. In the gray matter, neurons, dendrites, axons and other biological tissue are densely packed at sub-micrometer scales (Fig. 6.6).

Regarding hardware implementations, CMOS technologies are currently applied to fabricate fine-resolution two-dimensional electrode arrays. A group led by Andreas Hierlemann has successfully stimulated and recorded cultured neuronal networks

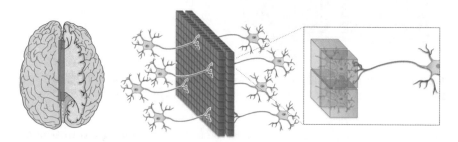

Fig. 6.5 The proposed BMI: a high-density double-sided two-dimensional array with a biomaterial coating for full read–write access to interhemispheric axonal bundles

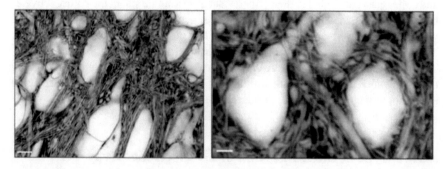

Fig. 6.6 Limitation of extracellular space due to high concentration of neurons, glia, dendrites, and axons. Scale bars are 4 μm (left) and 2 μm (right) (Reprinted from Tønnesen et al. 2018, Copyright 2018, with permission from Elsevier)

using a 17.5-μm pitch array with 26,400 channels. Once we have in hand two dimensional arrays with electrode spacing below a single micrometer, which is equivalent to the pitch distance of axons in the human corpus callosum, we will be able to independently record and stimulate all axonal fibers that integrate the two cortical hemispheres.

To materialize such a dream device, however, one technological development needs to be undertaken; the rescue of dissected axons. Generally speaking, if we leave dissected axons as-is, both ends will eventually die. We need to give up on the synaptic ends of axons, since they will not survive without an intact cell nucleus. As for the cell body side, however, if we can provide biological targets for dissected axons to reestablish connectivity, they will. Indeed, the capacity of axons to regrow and reconnect is applied in medical procedures, such as when curing complex bone fractures. Damaged motor axons will reach as far as centimeters to reconnect to their original muscle fiber. Therefore, if we embed biological tissue on the surface of an electrode array, it can serve as a target for dissected axons to reestablish connections. The wetware–hardware interface would then be far more robust and enduring, as compared to conventional methods that solely rely on extracellular electrodes to stay exactly in place and ward off tissue adhesion.

Although, I have to admit that such a wetware-hardware interface in the central nervous system requires a breakthrough. Unlike peripheral axons, axons in the central nervous system have little potential for regrowing and reconnecting. The bright side is that realizing such properties is a holy grail in the field of regenerative medicine, and significant resources are being applied to that goal.

Finally, let's consider how losing the synaptic side of dissected axons affects the brain-machine interaction. We would be relying on "antidromic" stimulation, say, action potentials running opposite the normal direction along axons. This might at first appear to be a critical limitation, since it deviates from the natural way of communication between the two biological hemispheres. Luckily, the majority of biological interhemispheric connectivity is "symmetric," with the sending layer within the six-layered laminar cortical structure in one hemisphere basically matching that of the receiving layer in the other hemisphere, largely negating this restriction. Simply put, antidromic stimulation via the proposed BMI would roughly follow the original format of interhemispheric communication. In fact, we cannot avoid antidromic stimulation anyway, regardless of where we place the electrode; axons in the vicinity of the electrode tip will always be activated, sending action potentials back to the soma. We may even claim that the proposed BMI is more tolerant to the artificiality of antidromic stimulation, since it only stimulates axons that were originally adopted for interhemispheric communication, while electrodes in the gray matter cannot control their axonal targets that accidentally run nearby.

Integration Between Artificial and Biological Hemispheres

Now that we have a BMI and a neutrally conscious machine potentially in place, let's move on to the second step of mind uploading: connecting and integrating our own consciousness with the machine. The question we want to address here is whether an artificial neural stream of consciousness will ever integrate with our private stream of consciousness.

We'll assume an artificial spiking neural network that replicates the full binary connectivity of a human cortical hemisphere. Regarding the quantitative values of synaptic efficacy, it would be acquired through a process of learning using general purpose sensory-motor material. Here, we assume that our BMI has the capacity to reproduce all interhemispheric neuronal connectivity in terms of quantity. Would the stream of consciousness in the mechanical hemisphere be integrated with the stream of consciousness in our biological hemisphere? Would we ever experience the machine's field of view?

At first glance, this state may appear similar to an intermediate stage of the digital fading qualia experiment. We would be at the step where we had started taking in neurons from one hemisphere, and had just completed reading in exactly half of the brain. In this context, it would be quite natural to assume that the two streams of consciousness would be integrated.

However, this overlooks an important point. Namely, preparing a machine hemisphere in advance and connecting it to a biological hemisphere is not equivalent to a halfway product of the digital fading qualia experiment. It skips the laborious process of taking in neurons one by one while perfectly recreating all its connections to the rest of the brain.

In this sense, the machine hemisphere that we have prepared is nothing like the original biological hemisphere. What are the chances of this machine hemisphere integrating, despite its differences from the original?

A key point here is that, even in the case of two biological brain hemispheres, neither has access to the fine anatomical structures and functional elements of the other. The only information that passes between them is action potentials, and there is precious little of that. All one hemisphere can access is a tiny glimpse of the other through the connections of the corpus callosum and the anterior and posterior commissures. The total number of interhemispheric axonal projections amounts to around 350 million, whereas there are 15 billion neurons in the human cerebral cortex. Interhemispheric connection exposes a mere 2% of the neurons to the other, and yet, that is all it takes to merge the consciousnesses of two biological hemispheres into one.

One point to consider, however, is that in the case of biological brain hemispheres, the two have been paired since they first developed in a fetus. Then again, the degree of freedom of the mechanical hemisphere and the BMI would compensate for the lack of history. For example, we could monitor what is visually presented to the biological hemisphere, analyze the properties of neurons through BMI readings, and adjust the BMI in a way that optimizes interhemispheric connectivity. Under our working hypothesis that integration of visual consciousnesses only requires them to function as a single generative model, the above adjustments are rather straightforward. It should be adjusted so that the high-level symbolic neurons are redundantly connected, say, a "red apple" neuron in the mechanical hemisphere forms a connection to the "red apple" neuron in the biological hemisphere and vice versa. We might also expect that if the brain–machine connection were maintained for long enough, the biological half would acclimate itself to the mechanical half.

Generally speaking, I believe that a neutral consciousness of a machine, if any, has a good chance to integrate with our biological consciousness. At the very least, it would be worth the shot.

Waking Up in a Machine

Let's imagine that a conscious machine has been perfected and connected to our own consciousness. In other words, we have solved the hard problem of consciousness through analysis by synthesis. What would be required to take the next step and fully transplant human consciousness into the machine? What must we do to wake up in the machine, breathe a virtual sigh of relief, and think, "Looks like the uploading was a success!".

To begin answering this question, it is helpful to step back and ask, why do we wake up every morning convinced that we are still us? Excluding the periods when we are dreaming, our consciousness is perfectly extinguished during sleep. Strictly speaking, in terms of continuity, the "I" of today is not the "I" of yesterday.

Nonetheless, we consider ourselves to be us because we retain memories. We confirm our continued existence through our personal history, from our earliest childhood memories up to what we had for dinner last night. We further support our continued personal existence through skills we have obtained—a foreign language we have learned, a sport we are good at, a musical instrument we may play.

Another important factor in defining ourselves is the way we make situational judgments. Even if we do not possess conscious free will, as Libet's experiments suggest, we each nonetheless have personal biases in the decisions that we subconsciously make. If we lost our habits in subconscious decision-making, distortions would arise in the process of antedated conscious reasoning, and we would surely lose a large chunk of our personal identities. These subconscious processes are also a type of memory; they are influenced and formed by past experiences.

If we were to wake up in a machine that retained all of our memories in the above sense, we would not pause and wonder who we are.

Transfer of Memory

We have arrived at the final third stage of the proposed mind uploading process; transfer of memory to the mechanical hemisphere, which Chalmers would categorize as an easy problem. (Fig. 6.7).

However, an easy problem in philosophy is not necessarily an easy problem in reality. Human brains contain tens of billions of complexly enmeshed neurons, and the totality of information contained within them resides in the delicate balance between thousands of trillions of synapses. Memories are fully integrated with the complex, enormous, highly detailed hardware of our brain.

Our memories are fully private, so we cannot resort to "neutral memories," as in the case of subjective experiences. Every event we experience during our lifetime shapes our memories via changes in synaptic connections, and those connections cannot be reproduced through a learning process using general-purpose sensory-motor material. We therefore need alternative methods for transferring our memories.

Storing memories after the integration of brain–machine consciousness is not much of a problem; we may replicate how the brain consolidates memory. The brain applies memory retention mechanisms that vary according to the type of memory. For instance, "episodic" memories—recollections of things like what you had for dinner last night and childhood events—are temporarily stored in a brain structure called the hippocampus. Those memories are "replayed" during sleep to reproduce cortical activity that accompanied the events during wakefulness. When these replays are repeated many times, Hebb's learning rule comes into play, consolidating memories in the cerebral cortex itself. If a similar mechanism is implemented in the mechanical

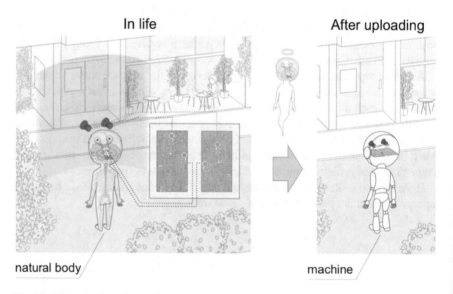

In life   After uploading

natural body   machine

Fig. 6.7 The transplantation process

hemisphere, then it is certainly conceivable that post-integration episodic memories can be safely stored on the machine side.

Transferring memories existing prior to machine–brain integration will be more challenging. Regarding episodic memory, a possible solution is resorting to active memory retrieval. Since consciousness would be readily integrated between the hemispheres, memory retrievals evoking cortical activity in the brain hemisphere would be translated to the mechanical hemisphere. More concretely, shared higher-level symbolic information will evoke generative processes of various modalities in the machine hemisphere, resulting in a constellation of neuronal activity in multiple areas. Then switching on the memory consolidation mechanism would do the rest.

The most problematic, perhaps, would be forgotten memories that we cannot actively recall, but nonetheless leave traces of synaptic connectivity that affect our subconscious decision making and emotional responses. Allow me to introduce an interesting set of clinical reports by the Canadian neurosurgeon Wilder Penfield that highlight such matters. During surgical procedures with local anesthetics, he exposed a large portion of the cerebral cortex, then electrically stimulated various points of the temporal lobe, where long-term memories are stored, and asked the patients to verbally report what they experienced. One such report went like this: "I had a dream. I had a book under my arm. I was talking to a man. The man was trying to reassure me not to worry about the book." Most of these experiences were genuine memories from the remote past, recalled in great detail, but the subjects claimed to have long forgotten them.

Taking advantage of Penfield's findings, we may transfer forgotten memories by applying electrical stimulation to the biological hemisphere via the BMI. As in the case of active recall, the replayed constellation of cortical neural activity could be translated over to the machine hemisphere and consolidated.

Bon Voyage!

I would like to close with a word to those readers who, like myself, are hoping to someday make the journey into the machine. Unlike other proposed methods, the transplantation process would be seamless. You would not need to die or have your brain extracted from your skull. It would be more like suffering a massive stroke in one of your hemispheres, where your consciousness would continue on in the intact hemisphere. Once your consciousness is fully integrated with the mechanical hemisphere, and sufficiently many memories are transferred to it, the closing of your biological hemisphere would be similar to a stroke, but in this case, your consciousness will seamlessly continue on in the machine.

We could even confirm intermediate transplantation checkpoints while your biological brain is still intact by turning off the BMI, waiting a few days, then reestablishing the connection. If you could recall events from the machine side as if they were from your own memory, that would confirm that part of you has safely resided in the machine. You could then be quite assured that your consciousness will continue within the machine, even after your biological brain comes to its inevitable end.

Afterword

In a letter to Robert Hooke, Isaac Newton once wrote, "If I have seen further it is by standing on the shoulders of giants."

Newton was known for his arrogance, so these were unusually modest words. The "giants" he referred to are, of course, his predecessors, who so painstakingly accumulated the scientific evidence on which he based his findings. Writing this book made me painfully aware of the extent to which modern-day neuroscience, too, stands on the shoulders of giants. As a researcher who only entered this field in the mid-1990s, I can hardly imagine neuroscience without Cajal's discovery of synapses, Hodgkin and Huxley's discoveries surrounding action potentials, Loewi's clarification of the role of neurotransmitters, Huber and Wiesel's observations of the stimulus-response characteristics of neurons, and many other key discoveries.

Considering the magnitude of these breakthroughs, it is certainly possible that the hard problem of consciousness will be solved, perhaps even sooner than we might expect, and that it will turn out to have not been so hard after all. Writing this book has made me hope all the more that such is the case.

The writing process has also forced me to put my thoughts into words to a greater extent than ever before, especially because my target audience is the general public. Previously vague concepts in my head became much more explicit and led me to new, interesting ideas. Honestly, when I started writing I intended this to be a much more somber book. I thought that its conclusions in particular would be more modest. But the more I think about it, the happier I am with how it turned out.

One lingering fear, however, is that I may have tricked you into reading a book that was not as "general interest" as you expected, and if that is the case, I extend my humblest apologies. I can only hope that any such tricks can be forgiven because through them you have gained a better appreciation for the depth of consciousness science. I hope you have seen how consciousness is one of the most wonder-filled of scientific fields, and that you, too, can sense the promise in our current faint glimmers of understanding.

Masataka Watanabe
March 2022

Appendix

The Generative Model in Its Simplest Form

To get a better idea of how the generative model works, let's take a look at the simplest model, which comprises only two visual areas. Local neural processing in a generative process is not like the telephone questionnaire we discussed in Chap. 2, but instead resembles a telephone tree, a pyramidal contact network used to rapidly distribute information among members of a group. Recall that the telephone questionnaire analogy focused on the role of neurons as receivers of action potentials. In contrast, the telephone tree analogy focuses on neurons' role as senders of action potentials and explains how they affect other neurons.

Let's look into a single iteration using a specific example. For simplicity, we will assume that lower visual area neurons here react to points. We could also assume that they react to lines, as demonstrated in the experiments by Hubel and Wiesel, but doing so complicates the illustrations, making it harder to see what is going on. We also assume that higher-level visual areas have neurons that react to complex objects such as "house" and "tree." "House" neurons fire when there is a house in the visual field, while "tree" neurons fire when a tree is visible.

Each neuron in these higher visual areas acts as the top distributor in the telephone tree, and is responsible for contacting multiple others in the network. To do so, dedicated telephone lines are assumed to link network members with those they are supposed to contact.

Imagine the head of this network, who will set off a pyramidal cascade of calls. Say that a "house" neuron fires. Then what we want is the neuronal firing pattern of a house in the lower level. The dedicated phone-line wiring needed to realize this is simple; the wires just need to end in the shape of a house (Fig. A.1). So when the higher-level "house" neuron fires, it sends action potentials to neurons in the lower-level area that are spatially arranged in the shape of a house. Given that the firing threshold for those neurons is low, they in turn produce a house-shaped firing pattern, as expected. We will also assume a similar configuration for the "tree" neuron, whose firing will result in a tree shape firing in the lower level.

© Springer Nature Switzerland AG 2022   157
M. Watanabe, *From Biological to Artificial Consciousness*, The Frontiers Collection,
https://doi.org/10.1007/978-3-030-91138-6

Fig. A.1 The generative process in a simple generative model

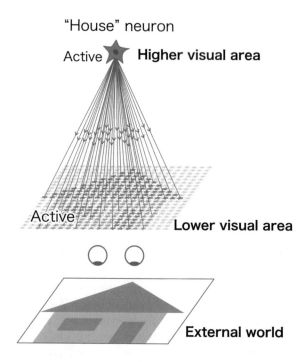

Calculating Generative Error

The lower level described above actually consists of three layers (Fig. 5.9). In addition to the lower-level generative layer, where the results of the generative process manifest, there is a layer that receives and reflects bottom–up sensory input, and another that calculates error between the two.

Let's see how this "generative error" is calculated within a single iterative process. To begin with, the sensory input layer directly conveys input such as light hitting the retina. Because we assumed that low-level visual areas respond to points, we can simplify the discussion by saying that a house in the visual field is reproduced as pointillistic neuronal firings in the shape of a house, and a tree will similarly produce a tree-shaped firing pattern (Fig. A.2).

The generative error refers to differences between firing patterns in the sensory input layer and the generative layer. In the example in Fig. A.2, there is both a house and a tree in the external world, but only the house neuron is firing in the higher level. There is thus a difference between the sensory input layer, in which both the house and the tree are represented as firing patterns, and the generative layer, in which there is only the house. Generative error is calculated by subtracting the firing pattern in the generative layer from the pattern in the sensory input layer. To perform this subtraction, the generative error layer is connected in parallel to the sensory input layer via positive synaptic connections, and also to the generative layer via negative synaptic connections (Fig. A.2).

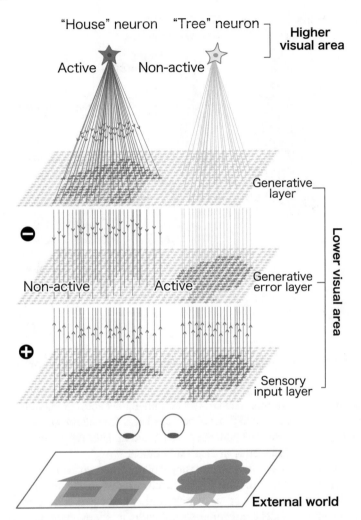

Fig. A.2 Calculation of generative error in a simple generative model

The final step in an iterative process is to use the calculated generative error to update activity patterns in the higher visual area to better match the external world.

The goal here is to determine which firing or non-firing of higher-level neurons caused the generative error. The model can easily determine the culprit if bottom–up connectivity from low-level generative error neurons to the high-level symbolic layer functions like the telephone questionnaire discussed in Chap. 2, the "tree" neuron questioning whether the generative error pattern resembles a tree, and the "house" neuron whether it more resembles a house (Fig. A.3). In this particular example, the generative error layer clearly shows a firing pattern in the shape of a tree, so the "tree" neuron would receive a multitude of positive synaptic inputs, causing it to fire in the

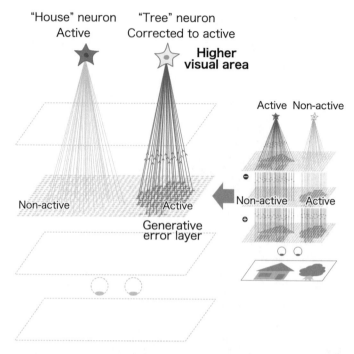

Fig. A.3 Generative error feedback to a higher-level site in a simple generative model

next iterative round. In this manner, the bottom–up synaptic projections from the generative error layer to the high-level neurons bears the pattern-matching function for detecting what caused the generative error, and updates the firing of high-level neurons to correct for that error in the next iteration.

This generative model is the objective target of my proposed natural law. It fits well with Revonsuo's assertion that the neural mechanism of consciousness is a virtual reality system shared between states of wakefulness and dreaming. During wakefulness, the generative model synchronizes itself with the external world using generative error based on sensory input. Generative error cannot be calculated during sleep, due to the lack of sensory input, and high-level symbolic representations are thus untethered from the restraints of the external world. Generative processes continue to work, however, so the virtual world of dreams remains generally consistent with what we experience during wakefulness.

Backward Propagation of Errors

However, there is one important element that cannot be properly captured in a generative model with only two levels, and I consider it a fundamental element for explaining our rich subjective experience. Allow me to set that aside for now, however, and first describe a certain mechanism that is required to incorporate this missing element into the generative model.

That mechanism is backpropagation of errors ("backpropagation" for short), a learning method at the core of the recent deep-learning boom in the field of artificial intelligence. Simply put, backpropagation is a learning rule that allows the training of neural networks comprising three or more layers. Backpropagation methods have been known since the 1960s, but they are finally experiencing their heyday thanks to a combination of faster computers, vast amounts of available training data, and innovative ideas.

Backpropagation has an interesting history. The discovery of the method is generally attributed to research performed in the mid-1980s by David Rumelhart, Geoffrey Hinton, and Ronald Williams. However, the concept was originally proposed two decades earlier by the preeminent Japanese theoretical neuroscientist Shun'ichi Amari.

Theoretical research on neural networks began soon after the Second World War, focusing primarily on learning mechanisms in a neural network comprising two layers. That changed in 1969, however, when a single book—Marvin Minsky and Seymour Papert's *Perceptrons*—exposed the theoretical limitations on the type of categories that two-layer neural networks can distinguish. This slowed research in the field overall, because at the time there were no known methods for training neural networks with three layers or more. Minsky and Papert later said that they wrote their book in hopes of spurring interest in theoretical neuroscience, but contrary to their expectations, researchers began leaving the field in droves, causing research on artificial neural networks to largely cease.

When a neural network has only two layers, training it is simple. If you want a given input to result in a particular output, delta-rule provides a clear method for varying the strength of synaptic efficacy in a way that will produce the desired output (Fig. A.4, top). To increase the output of a neuron in the output layer, you simply strengthen synapses from activated input neurons, so that they start providing larger input; to decrease output, you simply weaken the synapses from activated input neurons.

Things get much more complicated when a third layer is introduced, since it is not clear how we should modify neurons in the "hidden layer" between the initial and final layers. Amari proposed an elegant solution to this problem through use of a mathematical technique called partial derivatives, along with continuous input–output functions for individual neurons, instead of the traditional 0 or 1 step function.

Using this procedure, here is how the hidden layer modifies synaptic weights. A neuron in the hidden layer first takes a weighted average (according to its synaptic weights output to the final layer) of the final layer neurons' desired direction of change (the difference between the actual and desired output), and determines its own direction of change (Fig. A.4, bottom). Based on this calculated direction, it applies the delta learning rule to modify its synaptic weights arriving from initial-layer neurons. So in a sense, we can consider the error—the difference between the actual and desired output—as having traveled "backward" from the final layer to

Fig. A.4 The delta rule (top) and the backpropagation method (bottom) for machine learning

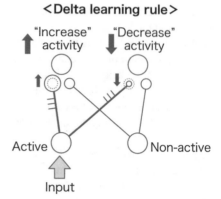

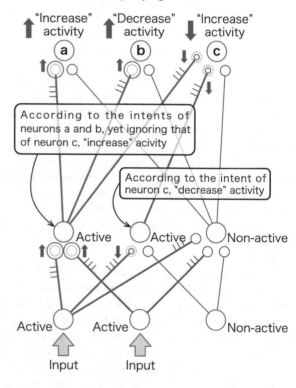

the hidden layer. This method was named error backpropagation two decades later, when Rumelhart and others rediscovered the same equation.

Creating Rich Middle Layers Through Learning

Learning through backpropagation in neural networks comprising three or more layers is also highly interesting from the perspective of understanding the fundamental nature of our brains. The initial layer corresponds to sensory nerves in the eyes, ears, and other sensory organs, and the final layer corresponds to the motor nerves by which the brain directly controls not only the muscles in our limbs, but also those used to move our eyes and produce sounds in our vocal cords. The key point here is that the brain is what connects those two layers, playing the role of hidden layers where the essence of neural processing is concentrated.

Getting back to the generative model, we have seen that only limited processing is possible in a model comprising two layers. In our simplest example, we saw that a firing of a "house" or "tree" neuron results in a flat firing pattern of a tree or a house in the generative layer. Of course, this simple mechanism is far from the three-dimensional visual processing that the brain realizes. At a minimal level, when one object is in front of another from the viewer's perspective, the front object should occlude what's behind it. A simple two-layer generative model, however, would only produce a semi-transparent combined image of the two objects. This is exactly the limitation that Minsky and Papert pointed out in *Perceptrons*.

To solve this occlusion problem, the higher level must add depth labels such as "in front of" and "behind" to its collection of symbolic information. Moreover, subsequent generative processes must occur through at least three neural network layers that require training through backpropagation.

In 2006, I worked on this occlusion problem along with Satohiro Tajima, then a student in my lab working on his bachelor thesis (see Box 1). No one at the time had predicted the rise of deep learning, so backpropagation was considered an antiquated technique, and our reviewers complained about our brazen use of it. I recall enjoyable discussions with Tajima on how to address such comments, and our deciding to press on at risk of rejection since there was no other method we could apply.

In summer of 2017, as I was writing the original Japanese version of this book, Tajima suddenly passed away. He had just taken an interest in problems related to consciousness, and published a groundbreaking paper. I cannot find the words to express my grief at the loss of his talent. He had an amazing proclivity for research from the time he was a student, and I had been looking forward to working with him again in the future.

Box 1: Applying Backpropagation in Generative Models

The serious limitations of generative models with just two layers preclude any chance of such a model being able to explain the three-dimensional virtual reality that resides in our brains. In 2011, pseudo-multilevel models created by piling up a number of two-level models had already been described in the literature, but these were unable to achieve nonlinear visual processing like

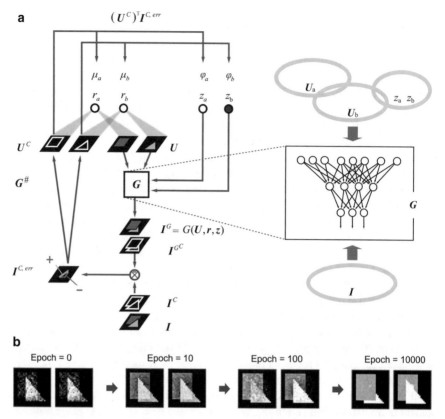

Fig. A.5 Using backpropagation to train an advanced neural network capable of occlusion (Adapted from Tajima and Watanabe 2011)

that required to obscure an object hidden behind another object. My student Satohiro Tajima and I therefore decided to attempt backpropagation training of a three-layer neural network to implement a generative process that can deal with occlusion.

Figure A.5 shows an overview of our model. It starts with a two-layer generative model like that described above, but replaces only its generative mechanism with a three-layer neural network. The problem then becomes how to train the system. When applying backpropagation to a visual-system model, one normally uses sensory input as system input, and trains the processing results for object recognition as output. However, the goal of the generative process is to create low-level representations from symbolic representations. We therefore decided to perform backpropagation training using symbolic representations as input and low-level representations as output. As symbolic representations, in addition to the presence or absence of visual objects (triangles and squares in this study), we also prepared neurons representing different depths in the field of view. As in normal generative models, values for symbolic representations were updated through generative error.

Before the training, two overlapping objects were inferred as being semi-transparent, like in a normal two-layer generative model. But with each training step, the firing pattern in the generative layer better reflected the correct occluded image.

Box 2: Odd Representations in Hidden Layers

One result of applying backpropagation methods is that predicting the characteristics of hidden-layer neurons becomes nearly impossible. As mentioned above, as the number of hidden layers increases, inferences pile onto inferences, thereby distributing information among many neurons. This produces odd response characteristics that obscure exactly what is being represented, and how.

In 1988, at the height of the second boom in research on artificial neural networks sparked by Rumelhart's rediscovery of backpropagation, David Zipser and Richard Anderson performed brain measurement experiments to ascertain whether hidden layers in the brain have similarly strange characteristics. They compared the results with those for hidden-layer neurons in an artificial neural network that was trained through backpropagation, a novel approach at the time. They conducted their neuron measurements in an area in the simian parietal lobe called 7a. This area is thought to use eye direction and the position of visual targets on the retina as inputs, from which—based on head position and orientation—it creates a coordinate system with the head at the origin and positions visual targets within those coordinates.

Figure A.6 shows an overview of the artificial neural network they used for the comparison, following the input–output relation in a simian 7a area. Back-propagation training produced proper positions as outputs, but the response characteristics of the hidden-layer neurons is particularly noteworthy. The figures at the bottom right of the figure show the response characteristics of two of those neurons. With eye orientation fixed, these three-dimensional graphs show the position of the visual target on the retina on the horizontal and depth axes, and the firing rate of the neuron on the vertical axis. Both neurons show the complex response characteristics that result from backprop-agation training, which results in hidden-layer information being distributed among many neurons. Interestingly, the response characteristics of the two measured 7a neurons (Fig. 5.15, bottom left) show similarly complex characteristics. Setting aside differences in the processes by which those neurons

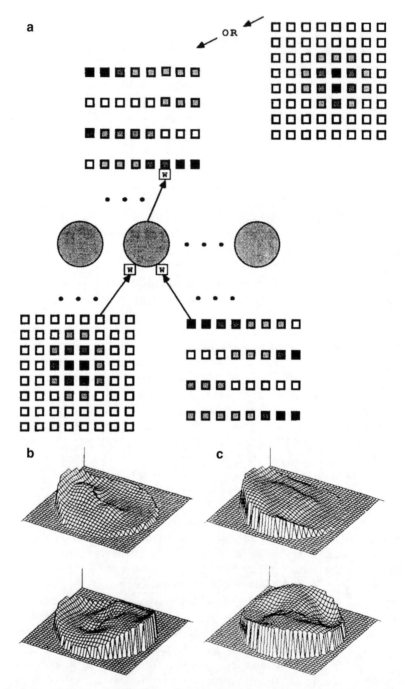

Fig. A.6 A comparison of neuron response characteristics in a simian parietal lobe and those in hidden-layer neurons simulated through backpropagation (Adapted from Zipser and Andersen 1988)

were trained, the end result was a brain neural network that is not very different from the artificial one trained through backpropagation

In truth, backpropagation could never occur in a brain exactly as described above. By its very name, backpropagation of errors implies propagating information in the reverse direction of its normal route, which in a biological brain would presumably be back up the axon from which it was delivered. Before the deep-learning boom, extensive research took place on training rules that could be implemented in a biological brain, but interest in such research has since waned; the limelight has been stolen by the reemergence of multi-layer neural networks as a method for machine learning, which was formerly cast aside purely for performance reasons. It is extremely interesting that an algorithm intended for modeling the brain has so extensively swept through the field of machine learning, where researchers are free to implement any model or method they can think up, without being tethered to concepts from biological systems. Considering how rapidly the machine learning field is advancing and how quickly algorithms demonstrating better performance replace those that came before them, I dearly wish I could ask some future readers of this book just how long backpropagation managed to retain its time in the spotlight.

References

Blake, R., & Logothetis, N. (2002). Visual competition. *Nature Reviews Neuroscience, 3*(1), 13–21.

Blakemore, C., & Cooper, G. F. (1970). Development of the brain depends on the visual environment. *Nature, 228*(5270), 477–478.

Chalmers, D. J. (1995). Absent qualia, fading qualia, dancing qualia. In T. Metzinger (Ed.), *Conscious experience*. Schöningh/ Imprint Academic.

Chalmers, D. J. (1996). *The conscious mind: In search of a fundamental theory*. Oxford University Press.

Cowey, A. (2005). The Ferrier Lecture 2004 what can transcranial magnetic stimulation tell us about how the brain works? *Philosophical Transactions of the Royal Society of London. Series b, Biological Sciences, 360*(1458), 1185–1205.

Cowey, A., & Stoerig, P. (1991). The neurobiology of blindsight. *Trends in Neurosciences, 14*(4), 140–145.

Crick, F., & Koch, C. (1990). Towards a neurobiological theory of consciousness. *Seminars in the Neurosciences, 2*, 263–275.

Crick, F., & Koch, C. (1995). Are we aware of neural activity in primary visual cortex? *Nature, 375*(6527), 121–123.

Džaja, D., Hladnik, A., Bičanić, I., Baković, M., & Petanjek, Z. (2014). Neocortical calretinin neurons in primates: Increase in proportion and microcircuitry structure. *Frontiers in Neuroanatomy, 8*, 103.

Edelman, G. M., & Tononi, G. (2000). *A universe of consciousness: How matter becomes imagination*. Allen Lane.

Florin, E., Watanabe, M., & Logothetis, N. K. (2015). The role of subsecond neural events in spontaneous brain activity. *Current Opinion in Neurobiology, 32*, 24–30.

Fujita, I., Tanaka, K., Ito, M., & Cheng, K. (1992). Columns for visual features of objects in monkey inferotemporal cortex. *Nature, 360*(6402), 343–346.

Gazzaniga, M. S., Bogen, J. E., & Sperry, R. W. (1962). Some functional effects of sectioning the cerebral commissures in man. *Proceedings of the National Academy of Sciences of the United States of America, 48*(10), 1765–1769.

Geldard, F. A., & Sherrick, C. E. (1972). The cutaneous 'rabbit': A perceptual illusion. *Science, 178*(4057), 178–179.

Gross, C. G., Rodman, H. R., Gochin, P. M., & Colombo, M. (1993). Inferior temporal cortex as a pattern recognition device. In E. Baum (Ed.), *Computational learning and cognition: Proceedings of the third NEC research symposium* (pp. 44–73). SIAM.

Histed, M. H., Bonin, V., & Reid, R. C. (2009). Direct activation of sparse, distributed populations of cortical neurons by electrical microstimulation. *Neuron, 63*, 508–522.

Hodgkin, A. L., & Huxley, A. F. (1939). Action potentials recorded from inside a nerve fibre. *Nature, 144*(3651), 710–711.

© Springer Nature Switzerland AG 2022
M. Watanabe, *From Biological to Artificial Consciousness*, The Frontiers Collection,
https://doi.org/10.1007/978-3-030-91138-6

Hodgkin, A. L., & Huxley, A. F. (1952). A quantitative description of membrane current and its application to conduction and excitation in nerve. *Journal of Physiology, 117*(4), 500–544.

Hubel, D. (1982). Exploration of the primary visual cortex, 1955–78. *Nature, 299*, 515–524.

Hubel, D. H., & Wiesel, T. N. (1959). Receptive fields of single neurones in the cat's striate cortex. *Journal of Physiology, 148*(3), 574–591.

Hubel, D. H., & Wiesel, T. N. (1962). Receptive fields, binocular interaction and functional architecture in the cat's visual cortex. *Journal of Physiology, 160*(1), 106–154.

Jiang, Y., Lee, A., Chen, J., Ruta, V., Cadene, M., Chait, B. T., & MacKinnon, R. (2003). X-ray structure of a voltage-dependent K+ channel. *Nature, 423*(6935), 33–41.

Johansson, P., Hall, L., Sikström, S., & Olsson, A. (2005). Failure to detect mismatches between intention and outcome in a simple decision task. *Science, 310*(5745), 116–119.

Kamitani, Y., & Shimojo, S. (1999). Manifestation of scotomas created by transcranial magnetic stimulation of human visual cortex. *Nature Neuroscience, 2*(8), 767–771.

Kanai, R., & Watanabe, M. (2006). Visual onset expands subjective time. *Perception & Psychophysics, 68*(7), 1113–1123.

Kastner, S., Pinsk, M. A., De Weerd, P., Desimone, R., & Ungerleider, L. G. (1999). Increased activity in human visual cortex during directed attention in the absence of visual stimulation. *Neuron, 22*(4), 751–761.

Kawato, M., Hayakawa, H., & Inui, T. (1993). A forward-inverse optics model of reciprocal connections between visual cortical areas. *Network: Computation in Neural Systems, 4*(4), 415–422

Koch, C. (2004). *The quest for consciousness: A neurobiological approach.* Roberts & Co.

Koch, C., & Tsuchiya, N. (2007). Attention and consciousness: Two distinct brain processes. *Trends in Cognitive Sciences, 11*(1), 16–22.

Kuffler, S. W. (1953). Discharge patterns and functional organization of mammalian retina. *Journal of Neurophysiology, 16*(1), 37–68.

Lamme, V. A. F., & Roelfsema, P.R. (2000). The distinct modes of vision offered by feedforward and recurrent processing. *Trends in Neurosciences, 23*(11), 571–579.

Leopold, D. A., & Logothetis, N. K. (1996). Activity changes in early visual cortex reflect monkeys' percepts during binocular rivalry. *Nature, 379*(6565), 549–553.

Libet, B. (2004). *Mind time: The temporal factor in consciousness.* Harvard University Press.

Loewi, O. (1908). Über eine neue Funktion des Pankreas und ihre Beziehung zum Diabetes melitus. *Archiv Für Experimentelle Pathologie Und Pharmakologie, 59*(1), 83–94.

Logothetis, N. K. (1998). Single units and conscious vision. *Philosophical Transactions of the Royal Society of London. Series b, Biological Sciences, 353*(1377), 1801–1818.

Logothetis, N. K., Pauls, J., Augath, M., Trinath, T., & Oeltermann, A. (2001). Neurophysiological investigation of the basis of the fMRI signal. *Nature, 412*(6843), 150–157.

Ma, L. Q., Xu, K., Wong, T. T., Jiang, B. Y., & Hu, S. M. (2013). Change blindness images. *IEEE Transactions on Visualization and Computer Graphics, 19*(11), 1808–1819.

Maier, A., Logothetis, N. K., & Leopold, D. A. (2007). Context-dependent perceptual modulation of single neurons in primate visual cortex. *Proceedings of the National Academy of Sciences, 104*(13), 5620–5625.

Maier, A., Wilke, M., Aura, C., Zhu, C., Ye, F. Q., & Leopold, D. A. (2008). Divergence of fMRI and neural signals in V1 during perceptual suppression in the awake monkey. *Nature Neuroscience, 11*(10), 1193–1200.

Majima, K., Sukhanov, P., Horikawa, T., & Kamitani, Y. (2017). Position information encoded by population activity in hierarchical visual areas. *eNeuro, 4*(2).

Maruya, K., Watanabe, H., & Watanabe, M. (2008). Adaptation to invisible motion results in low-level but not high-level aftereffects. *Journal of Vision, 8*(11), 7, 1–11.

McMahon, D. B., Bondar, I. V., Afuwape, O. A., Ide, D. C., & Leopold, D. A. (2014). One month in the life of a neuron: longitudinal single-unit electrophysiology in the monkey visual system. *Journal of Neurophysiology, 112*, 1748–1762.

Mumford, D. (1992). On the computational architecture of the neocortex. II. The role of cortico-cortical loops. *Biological Cybernetics, 66*(3), 241–251.

Musk, E., & Neuralink (2019). An integrated brain-machine interface platform with thousands of channels. *Journal of Medical Internet Research, 21*(10), e16194.

Obaid, A., Hanna, M.-E., Wu, Y., Kollo, M., Racz, R., Angle, M. R., Müller, J., Brackbil, N., Wray, W., Franke, F., Chichilnisky, E. J., Hierlemann, A., Ding, J. B., Schaefer, A. T., & Melosh, N. A. (2020). Massively parallel microwire arrays integrated with CMOS chips for neural recording. *Science Advances, 6*, eaay 2789.

Otten, M., Pinto, Y., Paffen, C. L. E., Seth, A. K., & Kanai, R. (2017). The Uniformity Illusion. *Psychological Science, 28*(1), 56–68.

Pandya, D. N., Karol, E. A., & Heilbronn, D. (1971). The topographical distribution of inter-hemispheric projections in the corpus callosum of the rhesus monkey. *Brain Research, 32*(1), 31–43.

Pascual-Leone, A., & Walsh, V. (2001). Fast backprojections from the motion to the primary visual area necessary for visual awareness. *Science, 292*(5516), 510–512.

Payeur, A., Guerguiev, J., Zenke, F., Richards, B. A., & Naud, R. (2021). Burst-dependent synaptic plasticity can coordinate learning in hierarchical circuits. *Nature Neuroscience, 24*, 1010–1019.

Peters, A. (2007). Golgi, Cajal, and the fine structure of the nervous system. *Brain Research Reviews, 55*(2), 256–263.

Pitzalis, S., Galletti, C., Huang, R. S., Patria, F., Committeri, G., Galati, G., Fattori, P., & Sereno, M. I. (2006). Wide-field retinotopy defines human cortical visual area V6. *Journal of Neuroscience, 26*(30), 7962–7973.

Radford, A., Metz, L., & Chintala, S. (2016). *Unsupervised representation learning with deep convolutional generative adversarial networks.* ICLR.

Ramachandran, V. S., & Blakeslee, S. (1998). *Phantoms in the brain.* William Morrow.

Rao, R. P., & Ballard, D. H. (1999). Predictive coding in the visual cortex: A functional interpretation of some extra-classical receptive-field effects. *Nature Neuroscience, 2*(1), 79–87.

Revonsuo, A. (1995). Consciousness, dreams and virtual realities. *Philosophical Psychology, 8*(1), 35–58.

Risse, G. L., LeDoux, J., Springer, S. P., Wilson, D. H., & Gazzaniga, M. S. (1978). The anterior commissure in man: Functional variation in a multisensory system. *Neuropsychologia, 16*(1), 23–31.

Rorschach, H. (1921). Psychodiagnostik. Methodik und Ergebnisse eines wahrnehmungsdiagnostischen Experiments. (Deutenlassen von Zufallsformen), Ernst Bircher, Bern.

Schwiening, C. J. (2012). A brief historical perspective: Hodgkin and Huxley. *Journal of Physiology, 590*(11), 2571–2575.

Sugase, Y., Yamane, S., Ueno, S., & Kawano, K. (1999). Global and fine information encoded by single neurons in the temporal visual cortex. *Nature, 400*, 869–72.

Tajima, S., & Watanabe, M. (2011). Acquisition of nonlinear forward optics in generative models: two-stage 'downside-up' learning for occluded vision. *Neural Networks, 24*(2), 148–158.

Tanaka, K. (1996). Inferotemporal cortex and object vision. *Annual Review of Neuroscience, 19*, 109–139.

Tong, F., & Engel, S. A. (2001). Interocular rivalry revealed in the human cortical blind-spot representation. *Nature, 411*(6834), 195–199.

Tønnesen, J., Inavalli, V. V. G. K., Nägerl, U. V. (2018). Super-resolution imaging of the extracellular space in living brain tissue. *Cell, 172*, 1108–1121.

Tononi, G. (2012). *Phi: A Voyage from the Brain to the Soul.* Pantheon Books.

Tononi, G., & Edelman, G. M. (1998). Consciousness and complexity. *Science, 282*(5395), 1846–1851.

Tootell, R. B., Switkes, E., Silverman, M. S., & Hamilton, S. L. (1988). Functional anatomy of macaque striate cortex. II. Retinotopic organization. *Journal of Neuroscience, 8*(5), 1531–1568.

Tsuchiya, N., & Koch, C. (2005). Continuous flash suppression reduces negative afterimages. *Nature Neuroscience, 8*(8), 1096–1101.

Wang, L., Weng, X., & He, S. (2012). Perceptual grouping without awareness: superiority of Kanizsa triangle in breaking interocular suppression. *PLoS One, 7*(6), e40106.

Watanabe, M. (2014a). A Turing test for visual qualia: an experimental method to test various hypotheses on consciousness. Talk presented at Towards a Science of Consciousness 21–26 April 2014, Tucson: online abstract 124.

Watanabe, M. (2014b). Turing test for machine consciousness and the chaotic spatiotemporal fluctuation hypothesis. UC Berkeley Redwood Center for Theoretical Neuroscience (video: https://archive.org/details/Redwood_Center_2014_04_30_Masataka_Watanabe)

Watanabe, M., & Aihara, K. (1997). Chaos in neural networks composed of coincidence detector neurons. *Neural Networks, 10*(8), 1353–1359.

Watanabe, M., Bartels, A., Macke, J. H., Murayama, Y., & Logothetis, N. K. (2013). Temporal jitter of the BOLD signal reveals a reliable initial dip and improved spatial resolution. *Current Biology, 23*(21), 2146–2150.

Watanabe, M., Cheng, K., Murayama, Y., Ueno, K., Asamizuya, T., Tanaka, K., & Logothetis, N. (2011). Attention but not awareness modulates the BOLD signal in the human V1 during binocular suppression. *Science, 334*(6057), 829–831.

Watanabe, M., Loewe, S., Vaiceliunaite, A., Logothetis, N., Katzner, S., Busse, L. (2015). Behavioral and neural effects of visual masking and optogenetic V1 suppression in mice. *45th Annual Meeting of the Society for Neuroscience*, Chicago, IL, USA

Watanabe, M., Nakanishi, K., & Aihara, K. (2001). Solving the binding problem of the brain with bi-directional functional connectivity. *Neural Networks, 14*(4–5), 395–406.

Watanabe, M., Shinohara, S., & Shimojo, S. (2011). Mirror adaptation in sensory-motor simultaneity. *PLoS One, 6*(12), e28080.

Watanabe, M., Totah, N. K., Kaiser, K., Löwe, S., Logothetis, N. K. (2014). Visual backward masking in rats: A behavioral task for studying the neural mechanisms of visual awareness. *44th Annual Meeting of the Society for Neuroscience*, Washington, DC, USA

Wu, D. A., Kanai, R., & Shimojo, S. (2004). Vision: Steady-state misbinding of colour and motion. *Nature, 429*(6989), 262.

Zipser, D., & Andersen, R. A. (1988). A back-propagation programmed network that simulates response properties of a subset of posterior parietal neurons. *Nature, 331*(6158), 679–684.

Zuber, B., Nikonenko, I., Klauser, P., Muller, D., & Dubochet, J. (2005). The mammalian central nervous synaptic cleft contains a high density of periodically organized complexes. *PNAS, 102*(52), 19192–19197.

Printed in the United States
by Baker & Taylor Publisher Services